103.4°E

Atlas of Remote Sensing of the Wenchuan Earthquake

31°N

CAS-PROJECT TEAM OF REMOTE SENSING FOR WENCHUAN EARTHQUAKE

Editor-in-Chief Guo Huadong

Co-Editors Zhang Bing Wang Jiesheng Dong Qing

CRC Press
Taylor & Francis Group
Boca Raton London New York

CRC Press is an imprint of the
Taylor & Francis Group, an **informa** business

CRC Press
Taylor & Francis Group
6000 Broken Sound Parkway NW, Suite 300
Boca Raton, FL 33487-2742

ISBN-13: 978-1-138-11217-9 (pbk)
ISBN-13: 978-1-4398-1674-5 (hbk)

Library of Congress Cataloging-in-Publication Data

Atlas of remote sensing of the Wenchuan earthquake / Huadong Guo.
    p. cm.
  ISBN 978-1-4398-1674-5 (hardcover : alk. paper)
   1. Earthquakes--China--Wenchuan Xian (Sichuan Sheng)--Remote-sensing maps. 2. Wenchuan Xian (Sichuan Sheng, China)--Maps. 3. China--Remote-sensing maps. I. Guo, Huadong. II. Title.

G2308.W35C55A8 2010
363.34'95095138--dc22
2009021571

Visit the Taylor & Francis Web site at
http://www.taylorandfrancis.com

and the CRC Press Web site at
http://www.crcpress.com

PROJECT TEAM OF REMOTE SENSING FOR WENCHUAN EARTHQUAKE CHINESE ACADEMY OF SCIENCES

- Center for Earth Observation and Digital Earth, CAS

- Institute of Electronics, CAS

- Institute of Remote Sensing Applications, CAS

- Institute of Geographic Sciences and Natural Resources Research, CAS

- Institute of Geology and Geophysics, CAS

- Institute of Mountain Hazards and Environment, CAS

- Research Center for Eco-Environmental Sciences, CAS

PREFACE

On 12 May 2008, Wenchuan County in Sichuan experienced a severe earthquake of magnitude 8.0 on the Richter scale, causing heavy losses in human life and property. Gaining immediate information from the disaster area and obtaining first-hand data was exceedingly difficult when the tremor first hit, since the roads were blocked, communications went down, many secondary disasters occurred, and weather conditions were very bad. The lack of information was a serious problem for disaster rescue decision-making since urgent monitoring and disaster assessment were vital if scientific support was to be provided for the search and rescue operations.

The Chinese Academy of Sciences (CAS) gave full attention to its comprehensive remote sensing techniques and talented scientists by immediately establishing the Emergency Headquarter of Remote Sensing for Earthquake Resistance and Disaster Relief. The Project Team of Remote Sensing Monitoring and Assessment of the Wenchuan Earthquake was set up by experts from seven CAS institutes, including Center for Earth Observation and Digital Earth, Institute of Electronics, Institute of Remote Sensing Applications, Institute of Geographic Sciences and Natural Resources Research, Institute of Geology and Geophysics, Chengdu Institute of Mountain Hazards and Environment, and Research Center for Eco-Environmental Sciences. Thus, remote sensing of the quake disaster was extended on all fronts. A program was immediately initiated to acquire data from earth observation satellites and two high-altitude airplanes deployed for obtaining high-resolution optical and radar data. On-the-spot investigations in the impacted area were carried out promptly. Therefore, a unified, all-weather day and night monitoring grid was formed with space, air and in-situ observing capacities. Disaster analyzing personnel did not spare any efforts day-in and day-out, collecting, transferring, processing, interpreting remote sensing data and images, and promptly completing series analysis reports. Those reports were immediately submitted to the Headquarters for Earthquake Resistance and Disaster Relief of governments at national, ministerial and provincial levels. The dedicated work provided convincing evidence for decision-making.

Accumulated knowledge in science and technology is a treasured fundamental resource for future progress, and summarizing knowledge scientifically is an important part in scientific and technological activities. Based on the accumulated data and images collected by the Project Team during quake relief, and requested by international readers, Taylor & Francis Group published *Atlas of Remote Sensing for the Wenchuan Earthquake*. Throughout the *Atlas*, the original appearances of quake-hit areas have been reconstructed thanks to the robust information and the in-depth analysis of researchers. The *Atlas* exhibits the disaster from six aspects, including geological disasters, barrier lakes, collapsed buildings, damaged roads, destroyed farmland and forests, and demolished infrastructure. The *Atlas* also demonstrates that the Dujiangyan Irrigation Project, which has been standing for 2000 years, remains fully functioning and keeps the Chengdu Plain operating optimally even after encountering an 8.0 magnitude earthquake. I believe that the publication will be welcomed by all circles, and play an important role in Wenchuan re-construction and regional development. I hope the *Atlas* can also serve as an informative reference for earthquake research.

The Wenchuan earthquake has warned us that natural disasters pose great challenges that humankind must face together. Natural disasters have occurred frequently in recent years. Earthquakes, tsunamis, floods and hurricanes have claimed tens of thousands of lives and caused heavy economic losses, bringing great trauma to human society. Effectively preventing natural disasters and reducing losses to a minimum have become real and pressingly important issues. Scientific and technological innovations should provide powerful support for disaster prevention and reduction. Different types of natural disasters require different solutions, thus development of new technical approaches, methods, and facilities are in great demand. A comprehensive earth observation system has to be established to capture abnormal phenomena in real time and near real time. We must monitor closely the occurrence and development of potential disasters and improve disaster warning ability. We should pay much attention to meteorological, geological, and seismological disasters that could cause serious damage to society. We will develop scientific models and carry out research on disaster simulation to increase our disaster warning ability and make contributions to global disaster research.

President Hu Jintao has recently set tasks for the scientific community on disaster prevention and mitigation. He has pledged to accelerate the applications of remote sensing, geographic information systems, global positioning systems and information communication technology, and urged speeding up the transfer and integration of high technologies for disaster prevention and mitigation. He has also called for setting up a national platform to share information on disaster mitigation and risk management, and to improve disaster monitoring, warning, assessment, and emergency rescue systems at both national and local levels.

It is my sincere hope that scientists and engineers in fields of remote sensing and geosciences courageously and enthusiastically assume these important tasks, and carry out the glorious mission that our country has bestowed upon us. I recommend that professionals 1) focus on disaster mitigation issues, 2) actively explore and recognize disaster mechanism, 3) develop key technologies for disaster monitoring, warning and prevention, 4) offer effective solutions, 5) provide strong scientific support that will safeguard human life and property, and 6) make contributions worthy of our time and abilities.

Lu Yongxiang
Vice Chairman, Standing Committee of National People's Congress, China
President, Chinese Academy of Sciences
June, 2009

FOREWORD

Earth Observation has played an increasingly important role in natural disaster mitigation with the rapid development of multi-platform, multi-frequency, multi-mode, high-spatial, ultra-spectral and high-temporal resolution technological capabilities. In recent years, reduction and prevention of disaster has been identified as the highest priority among nine societal benefits outlined by Global Earth Observation System of Systems (GEOSS), participated in by more than 70 countries and over 40 international organizations. In addition, the International Charter on Disaster Reduction and Cooperation was set up by multi-nation space organizations to provide a unified system of space data acquisition and delivery when confronting major natural disasters. China has also attached great importance to the application of space technology for natural disaster relief.

Immediately after the Wenchuan earthquake on May 12, 2008, the Chinese Academy of Sciences (CAS) deployed its earthquake remote sensing resources, setting up the CAS Emergency Headquarter of Remote Sensing for Earthquake Resistance and Disaster Relief. The CAS Project Team of Remote Sensing Monitoring and Assessment of the Wenchuan Earthquake was established by researchers from seven CAS institutes. Afterwards, remote sensing on quake-hit areas was fully extended. Within a month, the Project Team had taken advantage of two mega-science facilities of the remote sensing satellite ground station and the high-altitude remote sensing airplanes to give full play to the strengths and abilities at their command. The Project Team provided large amounts of data, information and suggestions for decision-makers at central and local government level, following the procedure of data-acquiring, information processing, disaster analysis and assessment, and report submission. At the same time, co-organized by the Ministry of Science and Technology (MOST) and Chinese Academy of Sciences, a data sharing mechanism was formed at a meeting attended by representatives from 13 ministries, and consultation meetings hosted by the Center for Earth Observation and Digital Earth (CEODE) were organized.

Remote sensing technology has played an active role in the Wenchuan earthquake monitoring and assessment. Remote sensing data recorded lasting and irreplaceable instantaneous pre- and post-quake landscapes. The purpose of the Atlas of Remote Sensing for the Wenchuan Earthquake is to systematically summarize the scientific results, promote academic exchange, and popularize remote sensing knowledge for disaster monitoring as well as to realize data sharing in a larger scope.

The *Atlas* consists of eight chapters. The first chapter, "Remote Sensing Data" introduces optical and radar data on quake-hit areas acquired through airborne and spaceborne remote sensing. The second chapter, "Geological Disasters" describes disasters caused by landslides, avalanches, detritus flows and fractures. The third chapter, "Barrier Lakes" narrates the distribution and dangers of the barrier lakes, one of the most serious secondary disasters caused by the earthquake. The fourth chapter, "Collapsed Buildings" analyzes spatial distribution and degree of damage of crushed buildings in urban and rural areas. The fifth chapter, "Damaged Roads" illustrates the five-level damage classification for national and provincial highways, county roads, and rural paths. The sixth chapter, "Destroyed Farmlands and Forests" assesses damaged conditions of forest vegetation and farmland resulting from the secondary geological disasters caused by the earthquake. The seventh chapter, "Demolished Infrastructure" shows the damage to hydrological engineering systems, mining area constructions and power transmission projects. The eighth chapter, "Civilization Perseveres" depicts a two-thousand-year old Dujiangyan Irrigation Project, which has been functioning even after encountering an 8.0 magnitude earthquake.

After the earthquake, we carried out field investigations in quake-hit areas including Dujiangyan, Wenchuan, Beichuan, Mianyang, and Mianzhu to validate remote sensing results. Our experience has made us reflect deeply on disaster relief work. Firstly, an emergency disaster monitoring system with advanced, practical, fast and reliable technical ability should be further established. Secondly, in an emergency, an authoritative institution should be guaranteed with a "scientific dispatching system" so that it could effectively coordinate key space infrastructures to achieve a high degree of data sharing among different agencies. Thirdly, the Earth has existed for 4.5 billion years, and some natural events, regarded as disasters by human beings, were merely natural phenomena in the Earth's long-term evolutionary process. Therefore the existence of humankind is always in concomitance with natural calamities; we must wage our struggles with natural disasters in accordance with the idea that humankind should coexist harmoniously with nature.

During the earthquake resistance and disaster relief period, Ms. Liu Yandong, State Councilor; Prof. Lu Yongxiang, Vice Chairman of the Standing Committee of the National People's Congress and CAS President; Prof. Bai Chunli, CAS Executive Vice-President, and CAS Vice-Presidents Jiang Mianheng, Ding Zhongli,Yin Hejun and other CAS senior administrators as well as Deputy Minister Cao Jianlin of MOST came to the CAS Project Team to coordinate work and approve the Project. Our team colleagues worked shoulder by shoulder, day in and day out. During the period of compiling the *Atlas*, advisory committee of the *Atlas* headed by Prof. Chen Shupeng and Prof. Ma Zongjin offered scientific directions in various ways, and colleagues of the editorial board worked meticulously and conscientiously. On behalf of the Project Team and editorial board, I sincerely express my heartfelt thanks and extend my highest respects to everybody. The assistance of CRC Press, esppecially of Irma Shagla, Rob Loftis, Suzanne Lassandro, and who converted a draft English manuscript of the *Atlas* into finished publication, are gratefuly acknowledged. It is also with thanks to Anthony J. Lewis and Barbara A. Lewis, of Louisiana State University, who reviewed the manuscript in response to appeals for assistance in English editing. A number of associates assisted in this *Atlas*, and special thanks go to Gao Wei, Wang Changlin, Liu Jie, Ling Thompson, and Liu Chuansheng.

Guo Huadong
Center for Earth Observation and Digital Earth, Chinese Academy of Sciences
June, 2009

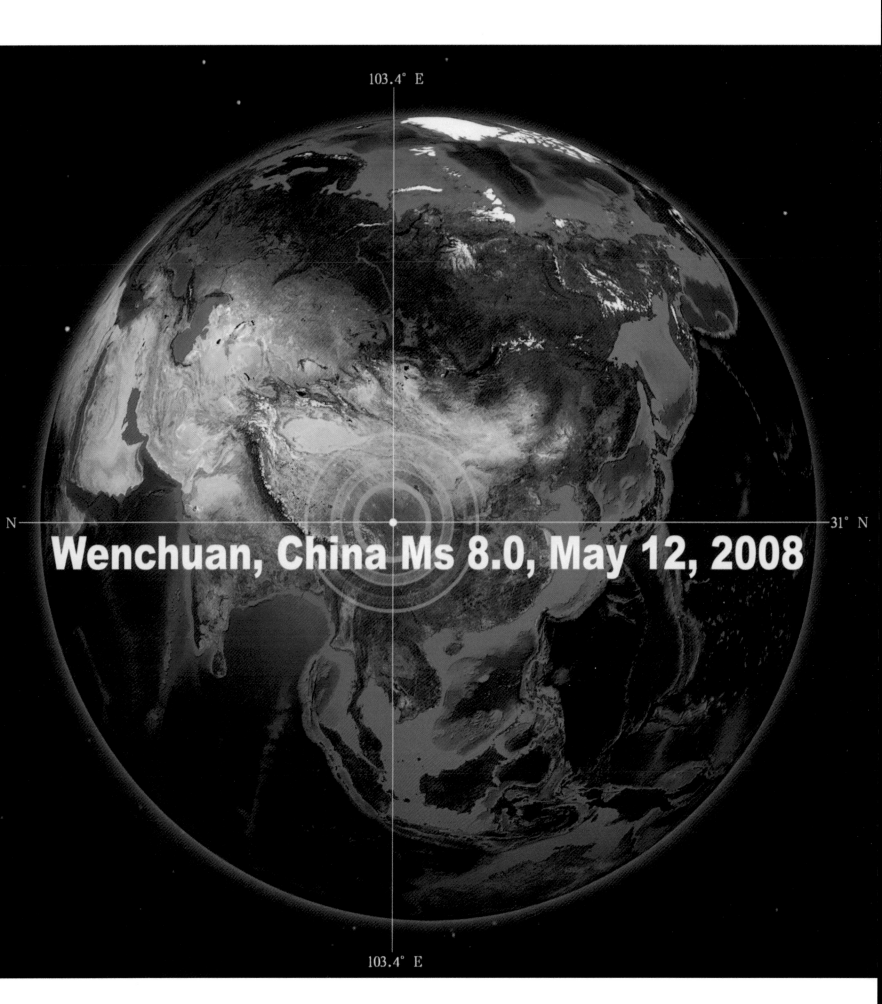

Wenchuan, China Ms 8.0, May 12, 2008

Distribution of Earthquakes over Ms 6 in the World

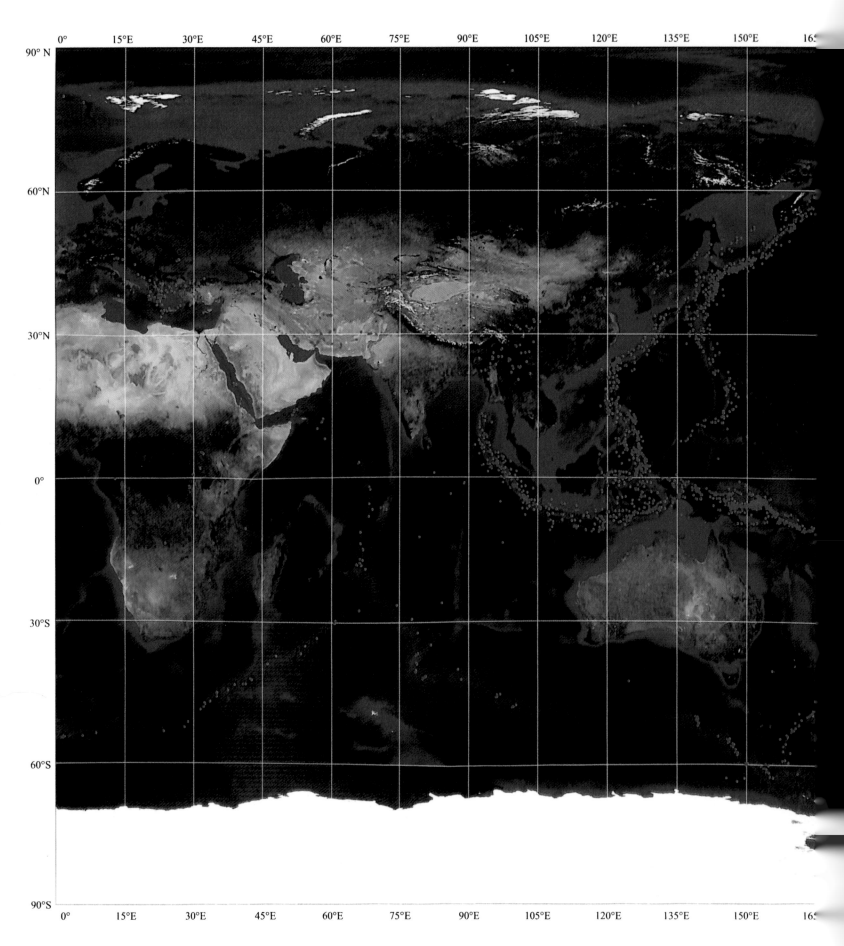

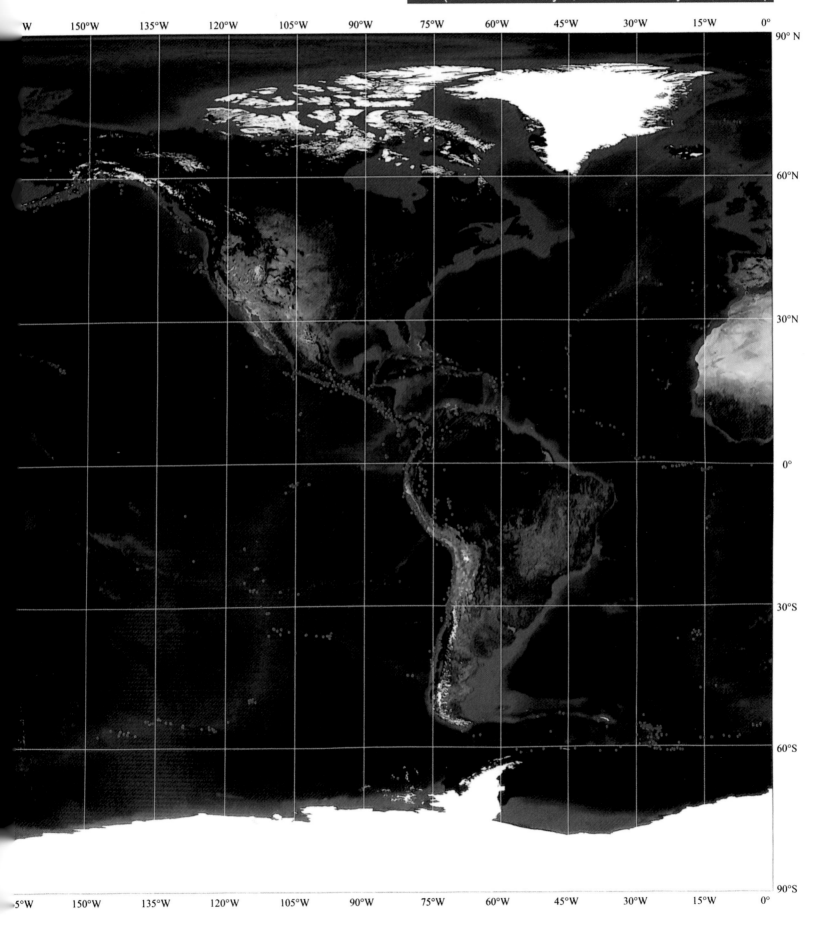

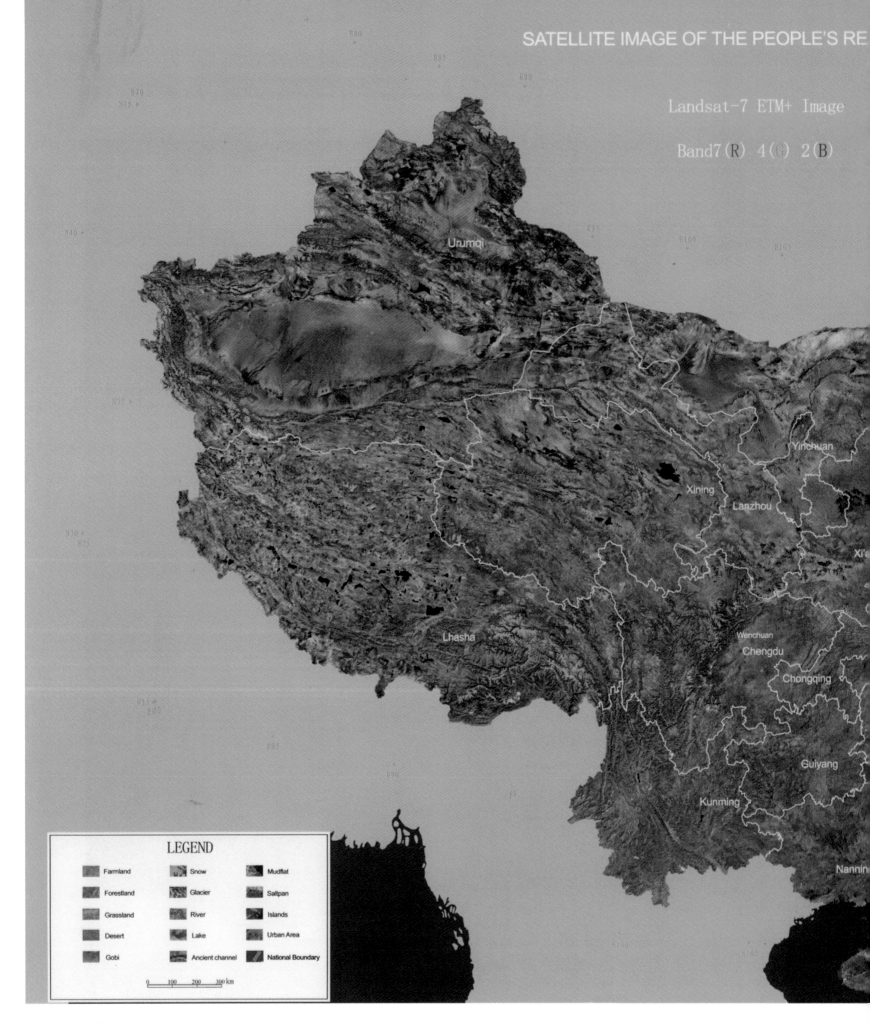

SATELLITE IMAGE OF THE PEOPLE'S RE

Landsat–7 ETM+ Image

Band7(R) 4(G) 2(B)

LEGEND

Farmland	Snow	Mudflat
Forestland	Glacier	Saltpan
Grassland	River	Islands
Desert	Lake	Urban Area
Gobi	Ancient channel	National Boundary

0   100   200   300 km

F CHINA

Harbin

Changchun

Shenyang

+ N40

Beijing

Tianjin

BOHAI (SEA)

Shijiazhuang

Jinan

ngzhou

YELLOW SEA

+ N35

+ N30

Hefei   Nanjing

Shanghai

Wuhan

Hangzhou

EAST CHINA SEA

Nanchang

sha

Fuzhou

Huangwei Yu Chihwei Yu
           Tiaoyu Tao
Pengchia Yu

+ N25

Taipei

Penghu Liehtao

Lü Tao
Lan Yu

Guangzhou

Hongkong

Macao

Dongsha Qundao

+ N20

E120

SOUTH CHINA SEA

Penghu
Liehtao

Lü Tao
Lan Yu

Dongsha Qundao

Xisha Qundao   Zhongsha Qundao

Huangyan Dao

SOUTH CHINA SEA

Nansha Qundao

Zengmu Ansha

0   200 km

Landsat Image

Bands: 7 (R) 4 (G) 2 (B)

Map Projection: Alber Conis

Equalarea Projection

Standard Parallels: 25° N, 47° N

Landsat image at the earthquake affected area

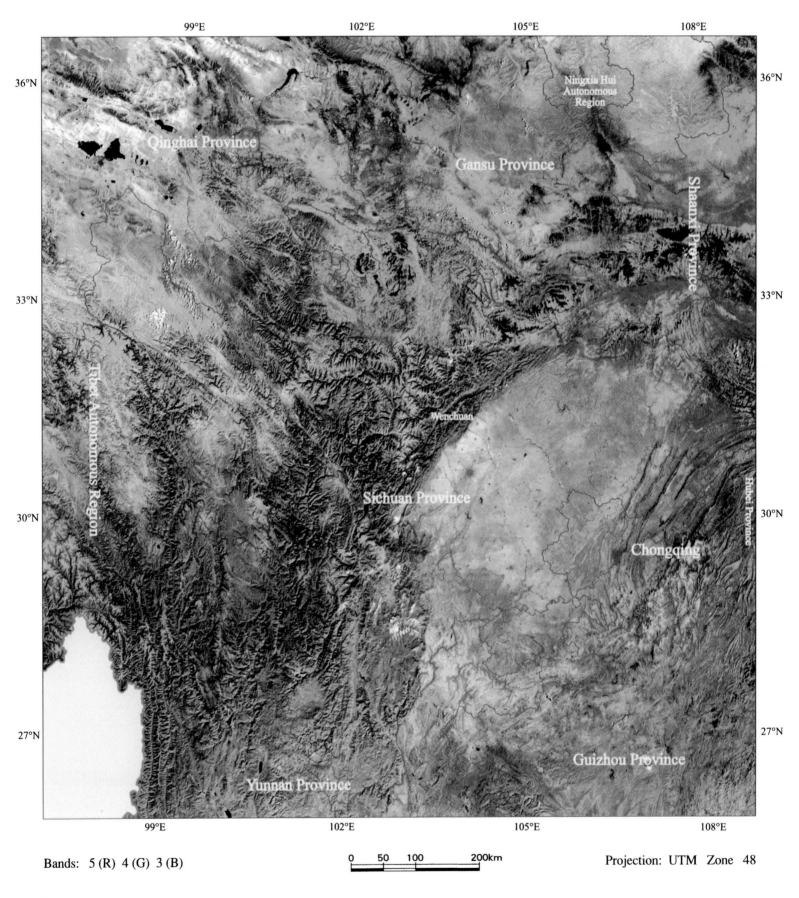

Bands: 5 (R) 4 (G) 3 (B)

0 50 100 200km

Projection: UTM Zone 48

· x ·

Map of Wenchuan earthquake area

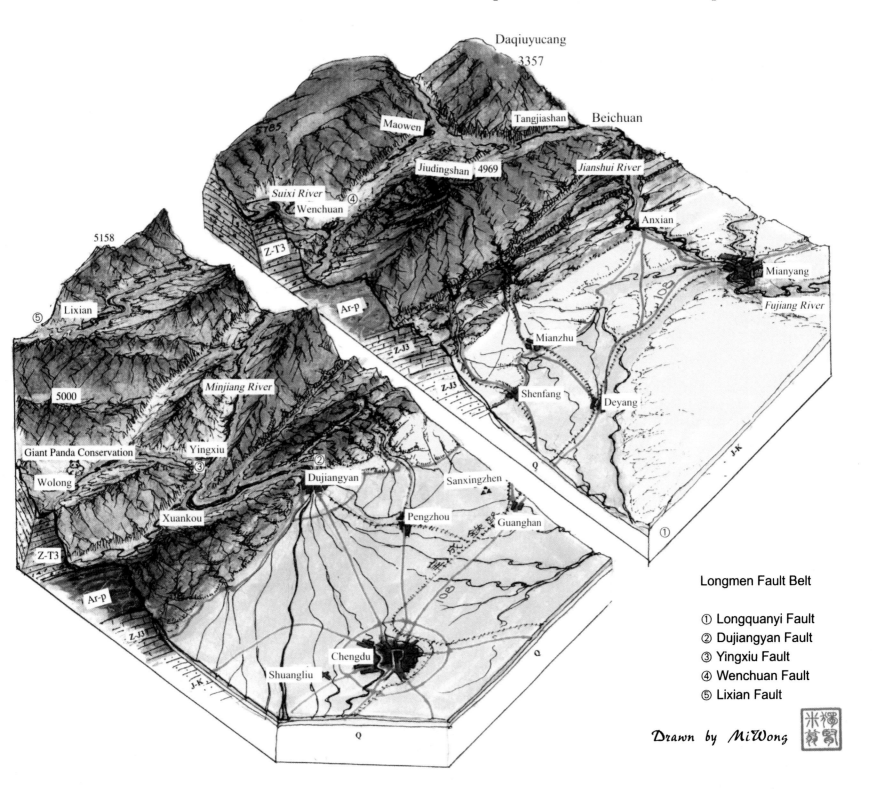

Daqiuyucang
3357

Tangjiashan   Beichuan

Maowen

Jiudingshan · 4969   *Jianshui River*

Suixi River   ④

Wenchuan   Anxian

Z-T3   Mianyang

Fujiang River

5158   Ar-p

⑤ Lixian   Z-J3

Mianzhu

Z-J3

Minjiang River

5000   Shenfang   Deyang

Giant Panda Conservation   Yingxiu

Wolong   ③   ②

Xuankou   Dujiangyan   Sanxingzhen

Z-T3   Pengzhou   Guanghan

Ar-p   Q

Z-J3

J-K

Longmen Fault Belt

Chengdu

Shuangliu   Q   ① Longquanyi Fault
② Dujiangyan Fault
③ Yingxiu Fault
④ Wenchuan Fault
⑤ Lixian Fault

J-K

Q   *Drawn by MiWong*

Three-dimensional image at the earthquake affected area

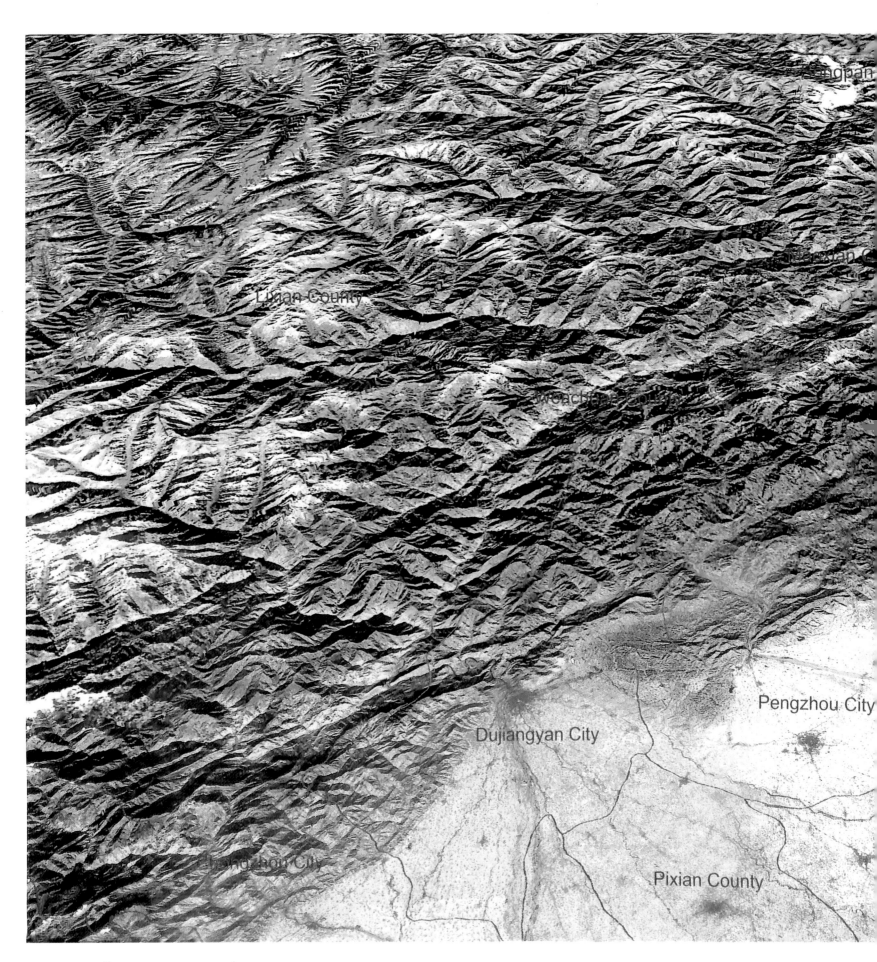

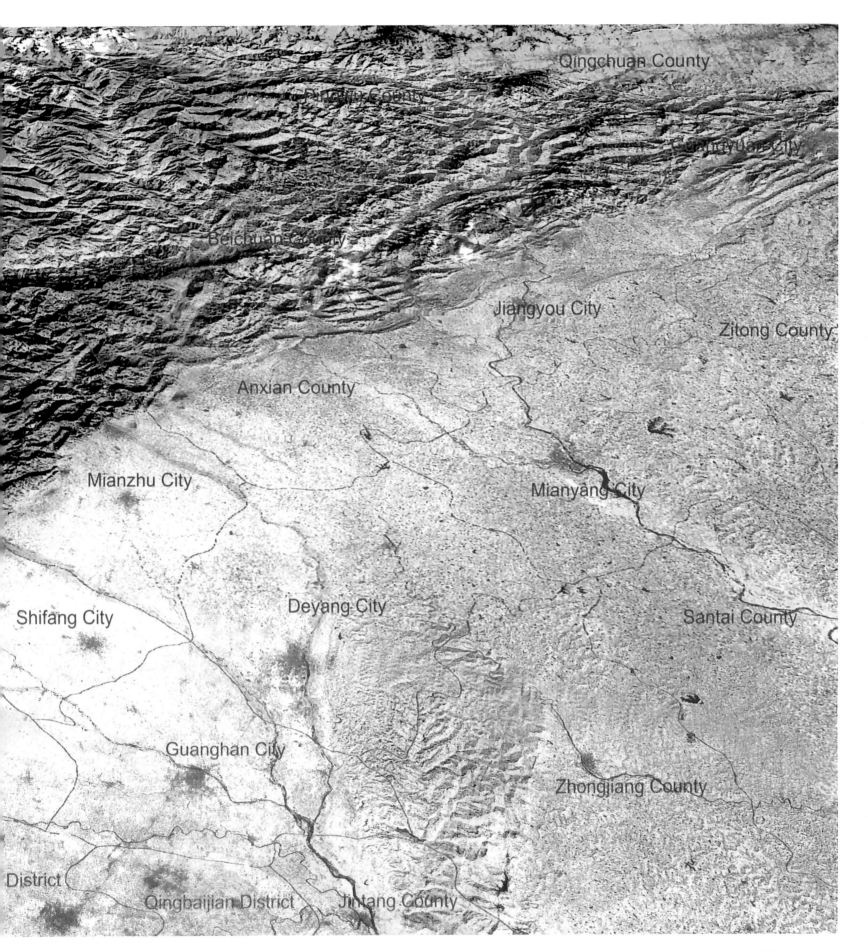

CONTENTS

Chapter 1 Remote Sensing Data

Chapter 2 Geological Disaster

Chapter 3 Barrier Lakes

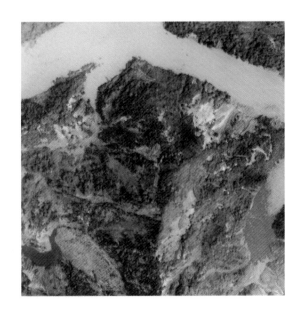

Chapter 4 Collapsed Buildings

Chapter 5 Damaged Roads

Chapter 6 Destroyed Farmlands and Forests

Chapter 7 Demolished Infrastructure

Chapter 8 Civilization Perseveres

Index

Chapter 1

REMOTE SENSING DATA

Remote sensing instruments can measure the characteristics of objects without contacting with them. Based on the principle that different objects have different responses to electromagnetic waves, remote sensing instruments collect data on the ground, detect the nature of surface objects, and identify various types of ground objects by sensors boarded on different platforms (tower, balloon, aircraft, rocket, manmade Earth satellite, spacecraft, space shuttle, etc.) above the ground. Remote sensing is the systematic integration of technologies for detecting and monitoring the Earth's resources and environment, including the process of detecting electromagnetic waves (radiation and scattering) on the Earth's surface, transmitting information, processing, and interpretation. After decades of development, remote sensing technology has been widely used in resource management, environmental studies, hydrology, meteorology, geology, urban studies, agriculture, forestry, surveying and mapping, disaster analysis, and other fields.

At present, the most commonly used remote sensing data acquisition platforms are satellites and aircraft. The United States' launch of the first Earth-observation satellite in 1972 marked the beginning of the era of space remote sensing. Since then, humankind has rediscovered the Earth from the perspective of space. Subsequently, the major powers developed a variety of photoelectric technologies; microwave technology and computer technology developed; and remote sensing technology entered a new phase when multi-resolution, multi-band, multi-polarization, multi-temporal massive Earth-observation data could be provided in real time. Satellite remote sensing has many benefits, such as wide coverage, repeated observation, and no restrictions from space.

Aerial remote sensing has matured as a technology since its beginnings in aerial photography reconnaissance. It has the advantages of allowing large-scale imaging, achieving high resolution, mapping across a wide terrain, and investigating detailed areas. However, there are limitations in flight altitude, endurance, attitude control, all-weather performance ability, and the range of dynamic monitoring. Aerial and satellite remote sensing could be used in conjunction, each performing applications to which it is suited.

Visible light, infrared light, and microwave are the bands usually used in remote sensing. Visible light works as camera. Infrared sensors are generally larger, are more complex in structure, and possess a more sophisticated system that can perceive the infrared band outside of visible light. The imaging mechanism of microwave remote sensing is different from that of the optical sensor, and synthetic aperture radar (SAR) is a common technology in microwave remote sensing. SAR emits electromagnetic wave energy, records backscattered signal strength from the ground, and displays an image after processing. Microwave is sensitive to dielectric constant and roughness of ground objects and is widely used in estimation of crop yield, forest surveys, disaster monitoring, marine applications, geological exploration, and military reconnaissance. Microwave technology does not rely on sunlight and has the ability to acquire images under all weather conditions and at all times, as well as having particular penetration capabilities. Microwave remote sensing data and optical data are highly complementary.

The satellite remote sensing data used in this *Atlas* are partly from the Center for Earth Observation and Digital Earth (CEODE) of the Chinese Academy of Sciences , others are provided by international organizations and companies. Aerial remote sensing images are obtained with two of CEODE's aerial remote sensing aircraft.

Satellite Remote Sensing Data

After the Wenchuan earthquake occurred, on the same day, CEODE immediately began to receive remote sensing data and access to foreign satellite data through international cooperation. Table 2.1 lists parameters about related satellite data used in this *Atlas*.

Table. 2.1. Remote Sensing Satellite Parameters

Satellite	Landsat-5	SPOT 2/4	SPOT 5	Resouresat	Envisat	Radarsat-1	ALOS	IKONOS	QuickBird	TerraSAR-X	EROS-B
Nation/ Institution	USA	France	France	India	ESA	Canada	Japan	USA	USA	German	Israel
Launch Time	1984	1990/1998	2002	2003	2002	1995	2006	1999	2001	2007	2006
Sensor Type	Optical	Optical	Optical	Optical	Radar	Radar	Optical /Radar	Optical	Optical	Radar	Optical
Sensor Name	TM	HRV	HRG /HRS	LISS3 /LISS4 /AwiFS	ASAR SAR		PRISM /Avnir-2 /PALSAR				
Orbital Type	Sun-syn-chronous	Sun-syn-chronous	Sun-syn-chronous	Sun-syn-chronous	Sun-syn-chronous	Helio-syn-chronous	Helio-syn-chronous	Helio-syn-chronous	Helio-syn-chronous	Helio-syn-chronous	Helio-syn-chronous
Orbital Altitude/km	705	832	832	778	800	798	691	681	450	514	510
Orbit Inclination/(°)	98.22	98.721	98.721	98.731	98.0	98.6	98.16	98.1	97.2	97.44	97.2
Orbital period/min	98.9	101.4	101.4	101.35	101	100.7	93.4	98.33	93.5	94.85	93.4
Imaging Method	Optical-mechanical scan	Push-broom	Push-broom	Push-broom	Synthetic aperture	Synthetic aperture	Synthetic aperture/ optical-mechanical scan	Push-broom	Push-broom	Synthetic aperture	Push-broom
Stereo-Imaging		Across-track	Along-track(HRS)/ across-track	Across-track	Across-track (Interferom-etry)	Across-track(Inte-rferometry)	Along-track	Along-track/ across-track	Along-track/ across-track	Across-track(Inter-ferometry)	Along-track/ across-track
Maximum Oblique viewing angle	27	27	26	15-45	10-49	44	60	30	20-55	45	
Swath (km)	185	60	60	23.9 /70 /141 /737	100/400	50 /75 /100 /150 /170 /300 /500	PRISM: 35/70 Avnir-2: 70 PALSAR: 20-350	11	16.5	10/30/100	7
Spatial Resolution (m)	30	10(Pan) 20 (Multisp-ectral)	2.5(Sper-mode) /5(pan)/10 (Multispec-tral)	LISS4: 5.8 LISS3:23.5 AWiFS: 56	25-100	10-100	PRISM:2.5 Avnir-2:10 PAL-SAR:7-100	0.82 (Pan) 3.2 (Multis-pectral)	0.61(Pan) 2.44(Mul-tispectral)	Spotlight:1 Stripmap:3 Scan-SAR:16	0.7(Pan)

▲ Sample image of *QuickBird* Satellite,Chengdu

The *QuickBird* optical satellite, which was launched in 2001 by Digital Globe, offered the most highly accurate and highest resolution imagery among the commercial satellites at that time. *QuickBird*'s global collection of panchromatic (resolution 0.61 m) and multispectral (resolution 2.44 m) imagery and color imagery, which is fused by the panchromatic and multispectral imagery, is designed to support applications such as map publishing, surveying, city design, agriculture and forest monitoring, digital information extraction, target recognition, and global information systems. During the Wenchuan earthquake, the *QuickBird* satellite data were used to determine appropriate methods and responses to dealing with different situations.

Table 2.2. *QuickBird* Primary Information

Launch date	October 18, 2001
Orbital altitude (km)	450
Revisit period (days)	1-3.5
Swath (km)	16.5
Resolution	0.61m(panchromatic), 2.44(multispectral)

Table 2.3. Technical Specification of Sensors in *QuickBird* Satellite

Resolution and Spectral Band	Panchromatic Ground sample distance (GSD) at nadir 0.61m
	Black and White:0.445-0.990 μ m
	Multispectral ground sample distance (GSD) at nadir:2.44m
	Blue:0.45-0.520 μ m
	Green:0.52-0.6 μ m
	Red:0.63-0.69 μ m
	Near-IR:0.76-0.9 μ m
Viewing Angle	Agile spacecraft-in-track and cross-track pointing
Per Orbit Collection	Approximately 57 single area images
Swath and Area Size	Nominal swath width:16.5 km at nadir; accessible ground swath:544 km centered on the satellite ground track
Metric accuracy	23 m circular error, 17 m linear error (without ground control)

◀ Sample image of *Landsat* 5 Satellite, Mianyang (May 19, 2006)

In the 1970s, America launched the first generation land resource satellite. In the 1980s, the second generation satellites (*Landsat 4, 5*) were in orbit too. There was a big breakthrough in this generation; and in the 1990s, the third generation satellites (*Landsat 6, 7*) were launched. Recently, *Landsat 5, 7* have been widely used over China.

Table 2.4. Technical Specifications of TM

Band	Spectral range (μ m)	Ground resolution (m)
B1 (Blue-Green)	0.45-0.52	30
B2 (Green)	0.52-0.60	30
B3 (Red)	0.63-0.69	30
B4 (Near IR)	0.76-0.90	30
B5 (SWIR)	1.55-1.75	30
B6 (LWIR)	10.40-12.5	120
B7 (SWIR)	2.08-2.35	30

▲ Sample image of *SPOT 5* Satellite (Mianyang, February 10, 2005)

The *SPOT 5* Satellite is representative of France's space remote sensing program. *SPOT 5* has higher resolution. It is the highest level of the *SPOT* satellites. It is used widely in agriculture, forest management, geology, urban planning, natural disaster management, and mapping.

Table 2.5. Specifications of *SPOT 5*

Launch Date	May 4, 2002
Orbital altitude (km)	832
Repeat cycle (days)	26
Swath (km)	60
Sensors	HRG/HRS

Table 2.6. HRG Spectral Bands of *SPOT 5*

Band	Spectral Range (μ m)	Resolution (m)
PAN	0.49-0.69	2.5/5
B1	0.459-0.61	10
B2	0.61-0.68	10
B3	0.78-0.89	10
SWIR	1.58-1.79	20

Resourcesat-1 (*IRS-P6*), launched in 2003 by India, operates three sensors: multi-spectral sensors LISS-4, LISS-3, and advanced wide-field scanner AWiFS. The highest resolution sensor is 5.8 m. The observed data are widely used in precision agriculture, forestry, environment, geological exploration, infrastructure planning, mapping, and national defense. The majority of the data applied to the May 12 Wenchuan earthquake relief were provided by LISS-4.

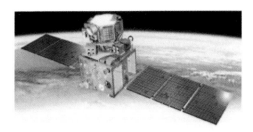

Table 2.7. Specifications of Indian *IRS-P6*

Date of Launch	October17, 2003
Mission lifetime (years)	5
Design Altitude(km)	817
Repeat cycle(LISS-III) (days)	24
Revisit period (LISS-IV) (days)	5
Repeat cycle(AWiFS) (days)	5

Table 2.8. Major Specifications of LISS-III

Items	Specifications
pixel number of CCD	6000
Band Range (μm)	B2(green): 0.52-0.59
	B3(red): 0.62-0.68
	B4(near infrared): 0.77-0.86
	B5(near infrared of short wave) 1.55-1.70
Swath (km)	141
Resolution(m)	23
Band registration accuracy	<0.25 pixel

▶ *Resoucesat-1*(*IRS-P6* LISS-3) sample image (SanXia area, March 8, 2008)

Table 2.9. Major Specifications of LISS-IV

Items	Specifications
pixel number of CCD	12,000
Band Range (μm)	B2(green): 0.52-0.59
	B3(red): 0.62-0.68
	B4(near infrared): 0.77-0.86
Swath(multispectral mode) (km)	23.9
Swath(Mono mode) (km)	70
Ground sampling distance (m)	5.8(nadir)
Side looking angle range	±26° (amount to ±398km on ground)
Band registration accuracy	<0.25 pixel

Table 2.10. Major Specifications of AWiFS Sensor

Items	Specifications
pixel number of CCD	6000
Band Range (μm)	B2(green): 0.52-0.59
	B3(red): 0.62-0.68
	B4(near infrared): 0.77-0.86
	B5(near infrared of short wave) 1.55-1.70
Swath (km)	740
Resolution (m)	56 (Nadir) , 7 (edge)
Band registration accuracy	<0.25 pixel

Launched in 1999 by a U.S. company, Space Imaging, *IKONOS* is the world's first commercial high-resolution satellite designed by Lockheed Martin. The spatial resolution is 0.8 m in panchromatic images and 4-m in multispectral mode. Panchromatic and multi-spectral images can be fused into the 1-m-resolution true-color mode. It opens up a faster and more economical way to obtain the latest Earth information and can partially replace aerial remote sensing. *IKONOS* is widely used in cities, ports, land, forest, environment, disaster investigation, and military monitoring. During May 12 Wenchuan earthquake relief efforts, the *IKONOS* images were applied for precise identification and interpretation of the disaster area.

Table 2.11. Primary Information of American *IKONOS*

Date of Launch	September 24, 1999
Size of satellite	Height,1.8m; Diameter, 1.6m
Orbit altitude (km)	681
Revisit period	1m, 2.9 days; 1.5m, 1.5days
Orbit type	Sun-synchronous
Weight (kg)	817 (1600 pounds)

Table 2.12. Major Specifications of Sensor

Band Range (μm)	Panchromatic: 0.45-0.90
	Muctispectral:
	B1(blue): 0.45-0.53
	B2(green): 0.52-0.61
	B3(red): 0.64-0.72
	B4(near infrared): 0.77-0.88
Swath (km)	11 (nadir)
Resolution	0.82 (panchromatic) 4 (muctispectral)

▶ *IKONOS* sample image (Wenchuan, May 23, 2008)

Launched in 1995 by Canada, *Radarsat-1* can work under all weather conditions. Its highest resolution is up to 10 m. *Radarsat-1* is used for both land and sea. Additionally, it was the first satellite to provide high-resolution data covering the whole Antarctic continent. With seven kinds of modes, 25 sorts of beams, and different angles of incidence, it provides images with 10-100 m multi-resolution and different swaths. *Radarsat-1* is widely used to monitor sea ice, analyze flooding, investigate oil pollution, and estimate agricultural yields.

Table 2.13. Major Specifications of Canadian *Radarsat-1*

Items	Specifications
Date of Launch	November 4, 1995
Design lifetime (years)	5
Orbit type	Sun-synchronous
Orbit altitude (km)	798
Revisit period(days)	24

Table 2.14. Technical Specifications of Canadian *Radarsat-1*

Instrument	Mode	Nomind Resolution (m)	Swath (km)	Incident Angle	Revisit Period (days)	Download Speed
SAR (Band C)	Fine	10	50	37° -48°	2-10	105 Mbps
	Standard	30	100	20° -49°	1-5	
	Wide	30	130-1652	0° -45°	1-5	
	ScanSAR narrow	50	300	20° -46°	1-5	
	ScanSAR Wide	100	500	20° -49°	1-5	
	Extended high	25	75	49° -59°	2-10	
	Extended low	35	170	10° -23°	1-5	

▶ *Radarsat-1* sample image (Dujiangyan, May 15, 2008)

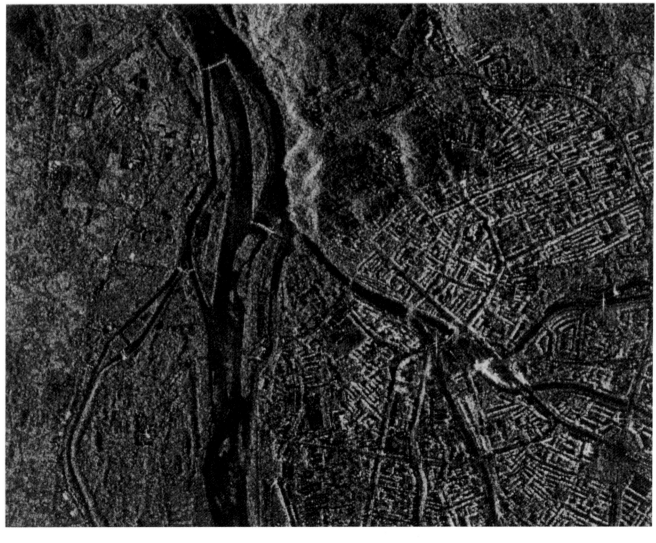

ALOS, a follow-up satellite of *JERS-1* and *ADEOS*, was successfully launched in 2006. It provides global high-resolution Earth-observation data with a resolution of up to 2.5 m. The satellite has three sensors: panchromatic remote sensing instrument for stereo mapping (PRISM), advanced visible and near infrared radiometer type 2 (AVNIR-2), and phased array type L-band synthetic aperture radar (PALSAR).

Table 2.15. Major Specifications of AVNIR-2

Items	Specification
numbers of Band	4
Wavelength (μm)	B1: 0.42-0.50
	B2: 0.52-0.60
	B3: 0.61-0.69
	B4: 0.76-0.89
Spatial Resolution(n)	10 (nadir)
Swath(km)	70 (nadir)

Table 2.16. Major Specifications of PRISM

Items	Specification
Number of Bands	1 (panchromatic)
Wavelength (μm)	0.52-0.77
Spatial Resolution (m)	2.5 (nadir)
Swath (km)	70 (nadir) 35 (triplet mode)

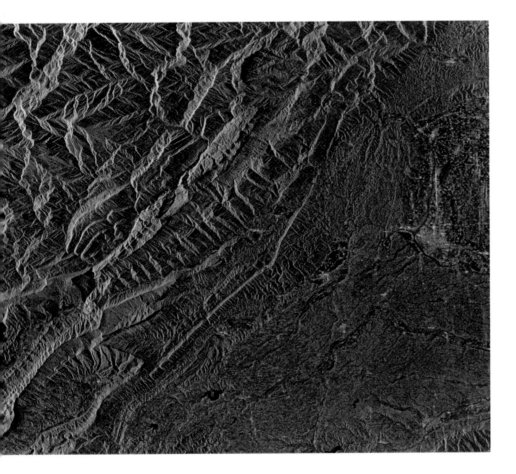

▲ *ALOS* PALSAR sample image (Dujiangyan, May 24, 2008)

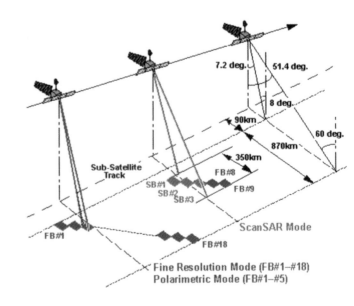

Table 2.17. Major Specifications of PALSAR

Items	Specification			
Mode	High resolution		ScanSAR	Polarization (experimental mode)
Center frequency	1270MHz(Band-L)			
Polarization mode	HH/VV	HH+HV/VV+VH	HH/VV	HH+HV+VH+VV
Incident angle	8°-60°	8°-60°	18°-43°	8°-30°
range resolution (m)	7-44	14-88	100 (multi-look)	24-89
Swath (km)	40-70	40-70	250-350	20-65

TerraSAR-X, a high-resolution radar satellite, was launched in 2007 by Germany. The satellite data spatial resolution is up to 1 m. It works under all weather conditions. It can penetrate vegetation to a certain extent, and can also obtain data similar to airborne images, and its polarization imaging mode can improve its object recognition ability. The data provided by *TerraSAR-X* are widely applied in national defense security, a variety of terrain thematic mapping, precision agriculture, urban and regional planning, environmental protection, coastal zone management and monitoring of ships, geology, disaster management, and other fields. The satellite data applied in the May 12 Wenchuan earthquake relief is mainly for precise interpretation of disaster areas.

Table 2.18. Specifications of German *TerraSAR-X*

Date of Launch	June 15, 2007
Orbit type	Sun-synchronous
Orbit Altitude (km)	505-533
Repeat cycle (days)	11

Table 2.19. Major Specifications of the SAR

Working modes	SpotLight	StripMap	ScanSAR
Wavelength	X-band (3.11cm)		
Polarization mode	VV/HH VV and HH	VV/HH VV+HH/HH+HV/VV+VH VV、HH、HV、VH	VV/HH
Spatial Resolution (m)	≤ 1	≤ 3	≤ 16
Swath (km)	10	30	100
Imaging direction	Right side of flying track		
Incident angle	20°-55°	20°-45°	

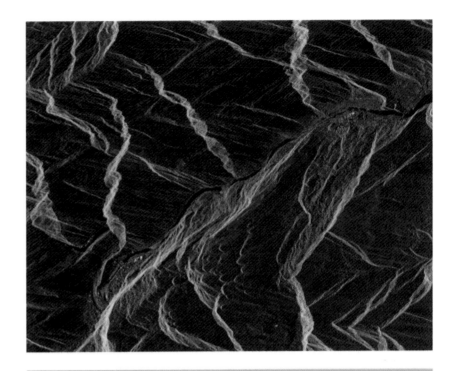

COSMO-SkyMed, a high-resolution radar satellite project, is the satellite jointly developed by the Italian Space Agency and the Italian Ministry of Defense. The satellite was launched in 2007 for the first time. The project consists of four X-band SAR satellites, the satellite data resolution is up to 1 m. The launch of the *COSMO-SkyMed* satellite constellation team will be completed in 2009. The main purpose of the project is to deal with dangerous situations, to monitor coastal zones, and to improve marine pollution in the Mediterranean surrounding areas. It is used for both military and civil purposes.

Table 2.20. *COSMO-SkyMed* Specifications

Items	Specifications
Polarization mode	HH、HV、HV、VH
Download channel	X-band
Compress ratio	6：3
Pulse repeat frequency (Hz)	3000
Bandwidth (MHz)	300
Swath (km)	30 (narrow band); 400 (wide band)
Range resolution (m)	3 (narrow band); 100 (wide band)
Azimuth resolution (m)	3 (narrow band); 100 (wide band)

◀ *COSMO-SkyMed* sample image (Beichuan, May 14, 2008)

Envisat, launched in 2002, is an important milestone in the Earth-observation satellite series developed by the European Space Agency. The satellite is the largest environmental satellite developed in Europe. It carries many sensors, which implement land observation, sea observation, and air observation. The most important of these sensors is an advanced synthetic aperture radar (ASAR). *Envisat* was applied to research of land, oceans, atmosphere, and so on, through continuous observation of cartography, to explore resources and to analyze global weather changes and natural disasters.

Table 2.21. Specifications of ESA *Envisat*

Date of Launch	October 17, 2003
Weight (kg)	8200
Designed lifetime (years)	5-10
Orbit Type	Sun- synchronous
Orbit Altitude (km)	800
Repeat cycle (days)	35

▲ *Envisat* sample image (Fujian, October 16, 2004)

Table 2.22. Major Specifications of ASAR

Working Frequency	C/5.3G				
Mode	image	alternating polarization	wide swath	global monitoring	wave
Spatial Resolution (m)	30	30	150	1000	10
swath (km)	≤100	≤100	About 400	About 400	5
Download speed(Mbps)	100	100	100	0.9	0.9
Polarization mode	VV or HH	VV/VH or HH/HV	VV or HH	VV or HH	W or HH

Airborne Remote Sensing Data

After the Wenchuan earthquake, the Center for Earth Observation and Digital Earth (CEODE), of the Chinese Academy of Sciences (CAS), sent two remote sensing aircraft into the disaster areas. From May 14, 2008 to June 8, 2008, the two remote sensing aircraft spent 227 hours in all to collect 23.7 terabytes (TB) data that include 18.5 TB radar data (X-band) of Institute of Electronics, CAS and 5.3 TB optical data (ADS40), which has created many records in the past 22 years.

▶ The two remote sensing aircrafts in the service of CEODE

Table 2.23. Technical Specifications of ADS40

Bands (μ m)	Panchromatic: 0.465-0.68
	Multispectral:
	B1 (Blue) : 0.43-0.49
	B2 (Green) : 0.535-0.585
	B3 (Red) : 0.61-0.66
	B4 (Near infrared) : 0.835-0.885
FOV	64°

Table 2.24. The Main Technical Specifications of Airborne X-band SAR

Working frequency (GHz)	X/9.6
Working bandwidth (MHz)	360
Zation polari mode	VV/HH/VH/HV
Mode	Strip imaging
Incident Angles	20°-70°
Beamuidth	2.3° (Azimuth), 12° (Range)
Resolution	0.5m × 0.5m

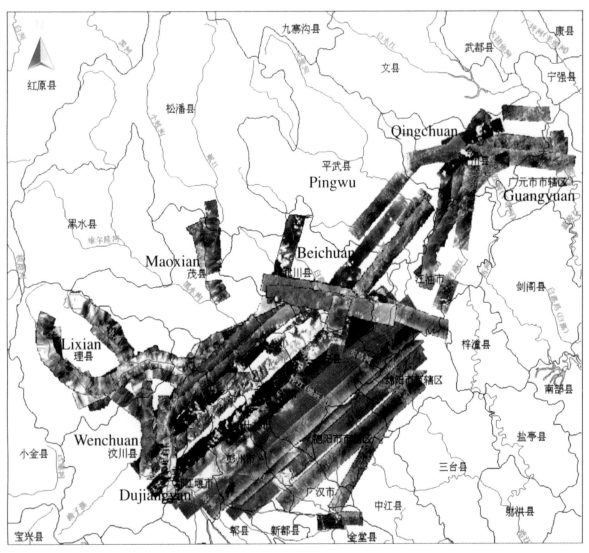

◄ Airborne optical data coverage of the Wenchuan earthquake area

▼ Airborne SAR data coverage of the Wenchuan earthquake area

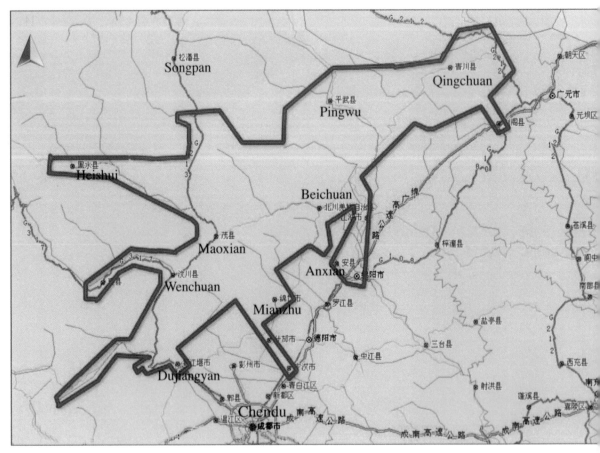

ADS40 image of earthquake-affected area

▲ Airborne X-band multi-polarization image in the suburban Dujiangyan, May 25, 2008

▼ CEODE has three ground stations locating in Miyun, Kashi, and Sanya, respectively. CEODE has the ability to receive and process remote sensing data in multi-station, multisatellite, multi-resolution, all-weather, all time, and quasi-real time conditions.

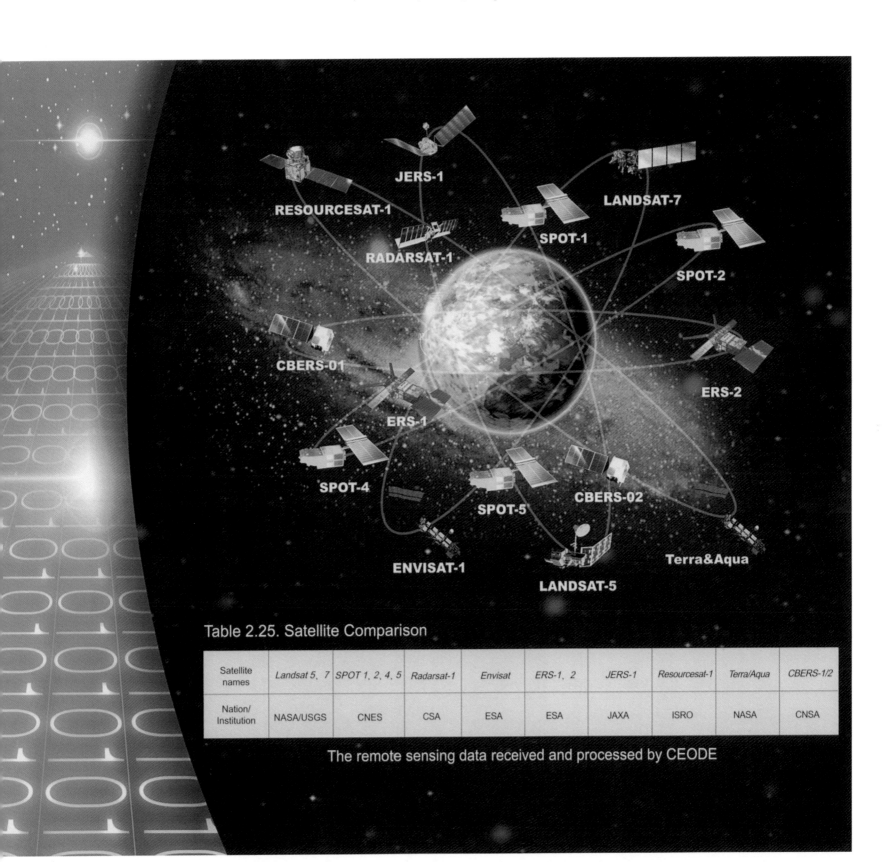

Table 2.25. Satellite Comparison

Satellite names	Landsat 5, 7	SPOT 1, 2, 4, 5	Radarsat-1	Envisat	ERS-1, 2	JERS-1	Resourcesat-1	Terra/Aqua	CBERS-1/2
Nation/ Institution	NASA/USGS	CNES	CSA	ESA	ESA	JAXA	ISRO	NASA	CNSA

The remote sensing data received and processed by CEODE

◀ The Miyun satellite ground station in suburban of Beijing

▲ The Kashi satellite ground station in Xinjiang Uygur Autonomous Region

▶ The design scheme for the Sanya satellite ground station in Hainan Province

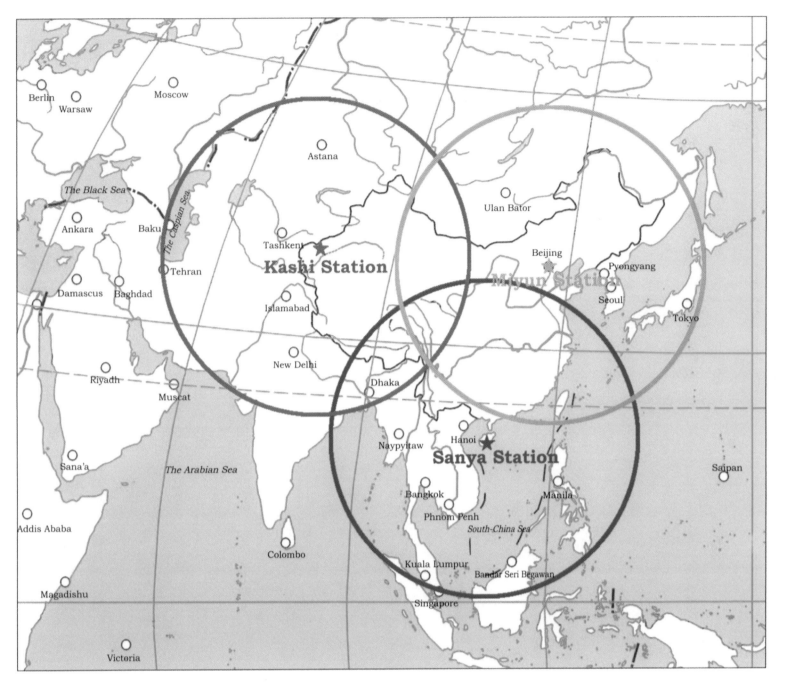

▲ The Miyun, Kashi, and Sanya ground stations can receive satellite remote sensing data covering 70% of Asian areas

Chapter 2

GEOLOGICAL DISASTER

Because of the geological and geographic conditions in Wenchuan, the Wenchuan earthquake triggered several secondary geological disasters, such as landslides, landslips, and debris flows, which not only led to heavy casualties and financial losses, but also severely damaged the roads, rivers, engineering projects, farmlands, and forests. The earthquake and the accompanying destruction of vegetation, landforms, rocks, soil, and other natural elements are the major causes of the change in the color and texture shown in the remote sensing images. The earthquake ruptured an approximately 29,000 km^2 area. The geological disaster covers an area of 2250 km^2, which is 7.8% of the total area.

The distribution of landslides, landslips, and debris flows can be described as strips along the river incision valleys and the areas with major engineering projects, which are mainly located in the valleys of the Minjiang River, the Jianjiang River, and their tributaries. The vertical elevation is from 800 m to 1500 m.

The quake affected area has been severely uplifted by tectonic movements and deep incised valleys formed, which have the characteristics of deep slopes, large catchments areas of valley, and wide distribution of quaternary stratum. The bulk of the geologic disasters occurred on slopes (mainly convex slopes) greater than 20 degrees, especially on slopes that are flat in the upper part and steep in the lower part and are close to the river and road. The phyllite and slate are both easily air-slaked. The hard terrain that causes the growth of crannies forms big rocks. The hard layers alternate with soft ones, which are more easily destroyed, and form loose materials consisting of not only big rocks but also debris. Because of gravity, all of the materials of air-slaked base rock accumulate at the slope, forming the origin materials of landslide and debris flow. Hard or half-hard layers, such as quartz sandstone and limestone, which are well fissured, tend to form landslips in the steep valley as a result of unloading. Potential landslide bodies bordered both sides of the valley. When the landslip rocks covered these potential landslides during the earthquake, the landslides lost balance and formed large landslip bodies. Because of the severity of the earthquake, the unstable slopes and potential landslide bodies formed landslides. In the Wenchuan earthquake, there were few single landslides. Because the rainstorm did not occur during the earthquake, there was no large mudslide, but more debris flow.

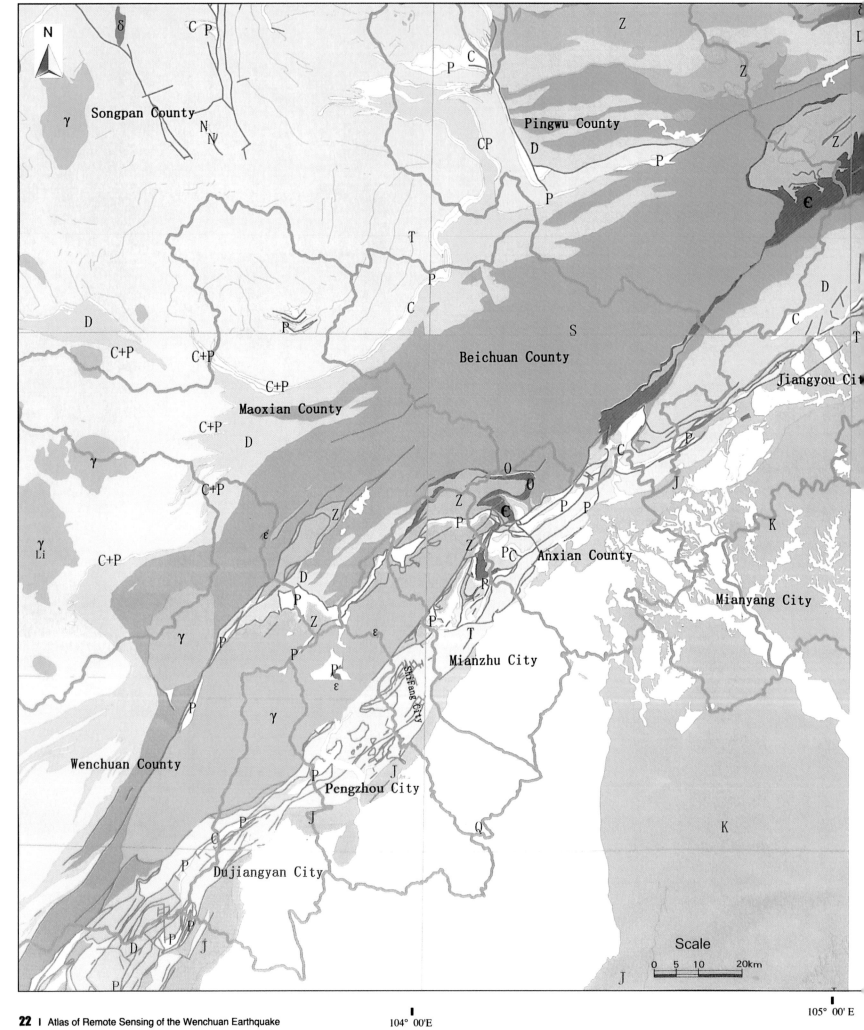

Songpan County

Pingwu County

Beichuan County

Jiangyou Ci

Maoxian County

Mianyang City

Anxian County

Mianzhu City

ShiFang City

Wenchuan County

Pengzhou City

Dujiangyan City

Scale

0 5 10 20km

104° 00'E

105° 00'E

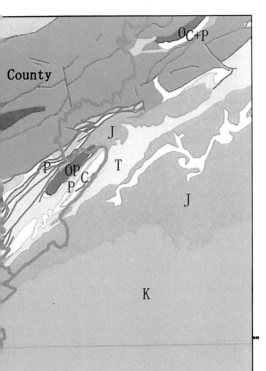

County

OC+P

J

P OP C
P

T

J

J

K

32° 00' N

Quate from 《1:200 000-scale
lodgital mapof china》. H-48-II
6), H-48-III (1971), H-48-VIII (1976),
8-XXVII (1970), I-48-XXXII (1975),
8-XXXIII (1971)。

Legend

═══ County

── Fault

▢ Q, River Heap

▢ N, Conglomerate Sandstone

▢ E, Conglomerate Sandstone

▢ K, Conglomerate Sandstone

▢ J, Sandstone Mudstone

▢ T, Sandstone Limestone

▢ P, Thick Limestone

▢ C, Limestone Marble

▢ C+P

▢ D, Quartz-Sandstone

▢ S, Sandstone Phyllite

▢ O, Limestone Marble

▢ Є, Conglomerate Sandstone

▢ Z, Sandstone Limestone

▢ Y, Granite Diorite

▢ δ, Diorite

▢ ε, Picrite

Regional Geological Map of Wenchuan Earthquake Area

The map shows structural limestone to the northeast. From southeast to northwest is the platform-edge sunken-break belt of Longmen Mountain, the Jia Lidong fold belt, and the Yansan fold belt. The folds are mainly closed line in shape and are often compound folds. The fault structures are often antifaults and anti-cover faults, which are always several kilometers to several hundred kilometers long. The Longmen Mountain break belt is mainly composed of the Guanxian-Jiangyou Fault (front range fault), the Yingxiu-Beichuan Fault (back range fault), the Wenchuan-Maoxian Fault (back range fault), and other relevant folds. The layers of fault zones are broken, and their occurrence is unpatterned; some are even backward. The break belt is accompanied by the base rock inversion. The break belts are several meters wide and have repetitious revival features. There are extrusion joints, mylonitize, and mylonite along the break belts.

Geology disaster distribution map of the Minjiang River

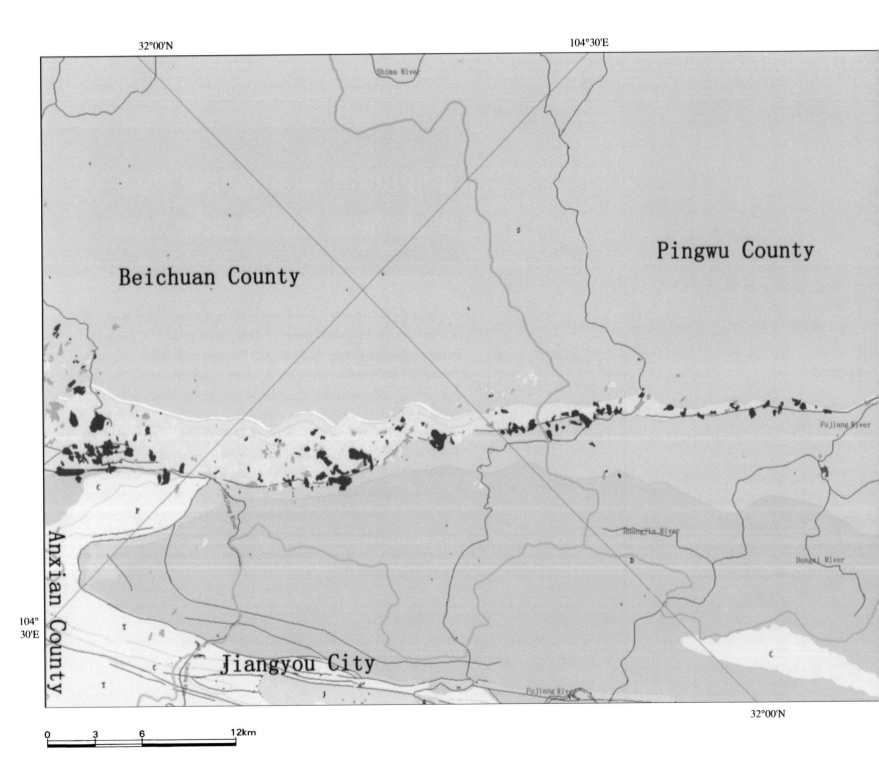

32°00'N   104°30'E

Shima River

Pingwu County

Beichuan County

Anxian County

104°
30'E

Fujiang River

HuangHa River

Dongxi River

Jiangyou City

Fujiang River

32°00'N

0   3   6   12km

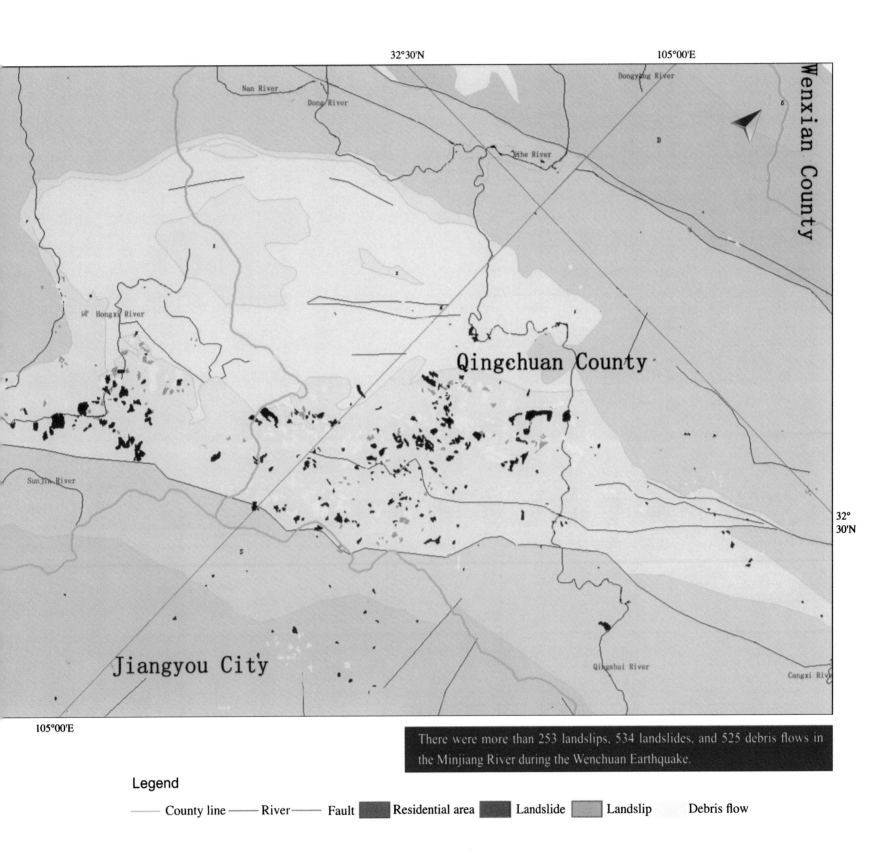

32°30'N 105°00'E

Wenxian County

Dongyang River

Nan River

Dong River

Yihe River

D

δ

Hongxi River

Qingehuan County

Sun Jia River

32°
30'N

S

Jiangyou City

Qingshui River

Cangxi Riv

105°00'E

There were more than 253 landslips, 534 landslides, and 525 debris flows in the Minjiang River during the Wenchuan Earthquake.

Legend

—— County line —— River —— Fault ▮ Residential area ▮ Landslide ▮ Landslip ▮ Debris flow

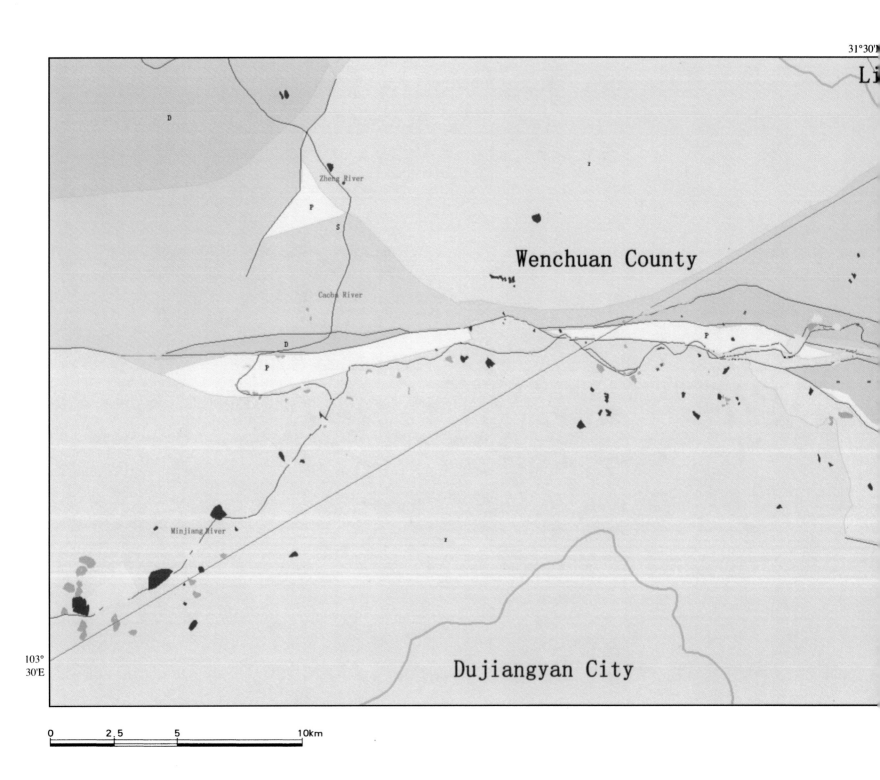

31°30'N

Li

D

Zheng River

P

S

Caoba River

Wenchuan County

D

P

P

Minjiang River

103°
30'E

Dujiangyan City

0   2.5   5   10km

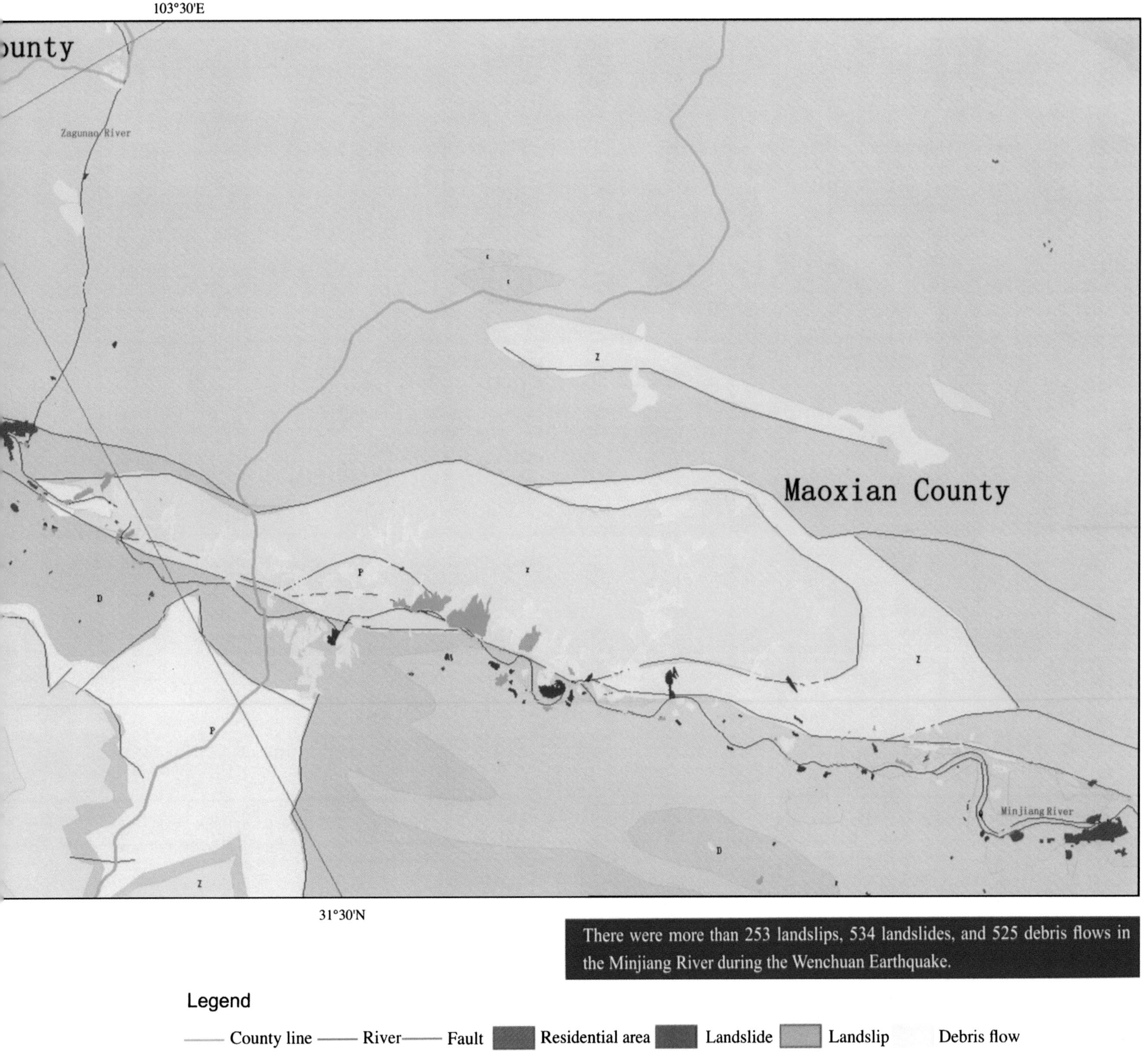
103°30'E
ounty
Zagunao River
Maoxian County
Minjiang River
31°30'N
Legend
County line
River
Fault
Residential area
Landslide
Landslip
Debris flow

There were more than 253 landslips, 534 landslides, and 525 debris flows in the Minjiang River during the Wenchuan Earthquake.

Yingxiu Town

▲ The photo shows a coseismic fault scarp with a height of ca. 2.5 m that has deformed the No.213 National Highway and formed a surface rupture zone as wide as 20 m (north-facing view).

◄ Aerial remote sensing image showing the earthquake fault around the town of Yingxiu, in Wenchuan County

The image shows irregular linear features of the earthquake faults that occurred on the second and the third terrace on the northwest bank of Minjiang River. Northwest of the town of Yingxiu, the earthquake faults appear as linear shallow tones (marked by arrows in the image) because of collapsed buildings along the fault zone. The earthquake fault traveled along National Highway 213 and the Minjiang River northeast of the town of Yingxiu. During uplift along the northwest wall of the fault (hanging wall), a swift current developed 200 m southeast of the Minjiang Bridge. On the northeast bank of the Minjiang River, the earthquake faults distributed along the east-northeast striking narrow valley.

Earthquake Fault

The surface rupture zones with a certain size and spatially continuous distribution produced by a large earthquake (about magnitude 6.5 and greater) are called earthquake faults. The surface rupture can reach and displace the ground surface, forming a fault scarp (steep break in slope). The resulting fault scarp may be several centimeters to 10 m in height, and up to about 430 km in length, depending on the size of the earthquake. The earthquake faults occurred along pre-existing active faults. Its distribution, occurrence, and characteristics of displacement are consistent with the active faults, and therefore become the indication of the active faults. Usually earthquakes above Ms 6.8 are associated with significant earthquake faults in China. Besides magnitude, the length and displacement (horizontal and vertical offset) of the earthquake faults are also closely related to the focal depth, the scale of the active faults, and the regional tectonic setting. The earthquake faults are characterized by the linear features displacing the hill ridge, streams, gullies, terraces, flood plains, and other landscape units on the ground surface in remote sensing imagery.

The May 12 Wenchuan earthquake ruptured two strands of the northeast-striking Longmen Shan Fault zone, producing a complex surface-rupture zone with a total length of 300 km. It is the longest surface-rupture zone from a thrusting event within a continental region ever reported. The major surface-rupture zone is dominated by thrust faulting, with up to 8–10 m of vertical displacement and 5.8 m of right-lateral offset.

▲ The photo shows an earthquake fault along the new dike of the Jianjiang River around the town of Qushan, in Beichuan County, and a surface-rupture zone with a width of 15 m formed along the surface-rupture zone (south-facing view)

▶ Aerial remote sensing image showing the earthquake fault around Qushan Town of Beichuan County

The northeast-striking earthquake fault passed through the town of Qushan, in Beichuan County, from the lower right side of the image. The fault's southwest section shows a light-yellow linear feature (indicated by the arrows in image) because of collapsed buildings along the earthquake fault. In addition, owing to the long-term effect of active faults, a landscape typical of meandering river features formed around the town of Qushan.

Beichuan

Pingtong Town

▲ The earthquake fault deformed a water pool and formed a coseismic fault scarp with a height of ca. 2.5 m, and the surface rupture zone is 15 m wide, observed from the river terrace west of the town of Pingtong (north-facing view)

◄ **Airborne remote sensing images showing the earthquake fault around the town of Pingtong, in Pingwu County**

In the lower left part of the image, the earthquake fault cut the river terrace and flood plain and created a fault scarp. Northeastward to the village of Lijiayuan, the earthquake fault destroyed most of the buildings, and it exhibited light-white linear features in the image. A waterfall formed along the southward flowing river (indicated by an arrow in the image) because of the uplift of the northwest wall of the earthquake fault. The surface-rupture zone appears as a linear image feature in the farmland on the east bank of the river. The earthquake fault passed through the streets of the town and extended toward the northeast.

► **Airborne remote sensing image showing the earthquake fault around the town of Shikan, in Pingwu County**

The earthquake fault cuts through the river terrace, producing fault scarp at the surface; it appears clearly as a linear feature in the image (indicated by arrows in the image). The earthquake fault terminates around the northwest of Shikan Town of Pingwu County.

▲ The surface rupture zone deformed the highway and formed a fault scarp ca.1.6 m in height with a right-lateral horizontal offset component, observed near the Shikan gasoline stop (northwest-facing view)

Shikan

Interferogram of SAR images obtained on February 17, 2008 and May 19, 2008. A large deformation along the center of the Longmen Mountain Fault (Yingxiu-Beichuan Fault) caused decorrelation of the interferogram. This deformation, together with the information of fault distribution from the ground survey, showed that most of the area had subsidence. The scale was approximately centimeters to meters, and there were two deformation centers within Beichuan County and the city of Mianzhu, the LOS deformation beyond 1 m. Since the rock area was stable with respect to the south plain area, the edges were relatively continuous, but the mountain area was located to the north of the fault. The edges were fractal because of the extrusion of the Tibetan Plateau.

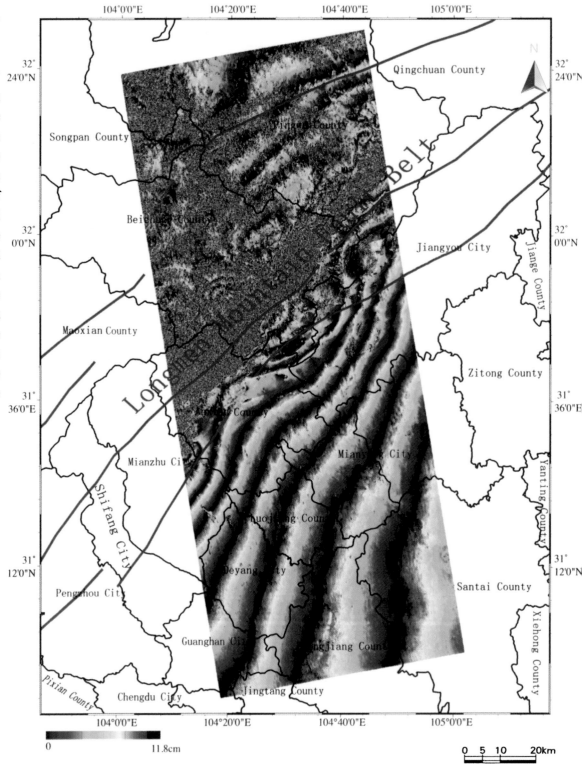

▶ **Airborne optical image of Nanxin Town in Maoxian County**

This image, obtained on May 28, 2008, shows large landslips near the LiangJiashan hill area, which caused damage to the vegetation and the side hill, where some parts of the land slid into the water. These events caused channel filling.

▶ **Airborne optical images of Lamamiao in Pingwu County**

This image, obtained on May 28, 2008, shows large landslips in the Lamamiao area; the slides dropped from above and piled up at the bottom of the area; the farmland and the trees were covered.

This image, obtained on May 23, 2008, shows large landslips near the town of NanXin. The slides followed along Minjiang River for about 4000 m and caused channel filling in some places along of the river. This endangered the road along the river, as well as water mains and electric lines.

▼ Airborne optical images of Matou in Wenchuan County

This image, obtained on May 16, 2008, shows three main landslips. The slide marked A is about 0.04 km² and may cause channel fill in the Minjiang River; the slide marked B is about 0.06 km² and has caused channel fill in the Minjiang River; the slide marked C is about 0.15 km². They are linearly distributed along the mountain slope and cause large areas of sloping and damaged the vegetation on the hill. This can evolve into mud-rock flow and cause new geologic hazards if there are aftershocks or heavy rains.

An image of debris flows caused by the earthquake in Minjiang River Basin

This image, obtained on May 16, 2008, shows large landslips close to the entrance of the Guxi Brook. The slide marked A has caused large areas of sloping on the hill and damaged the vegetation and farm field. Part of the road and house are destroyed. The slide marked B has caused vegetation damage and farm field damage.

▼ Airborne optical image of Jinhe phosphorite mine in the city of Mianzhu

This image, obtained on May 23, 2008, shows large landslips on the slope of the hill close to the Jinhe phosphorite mine. The slide is about 0.4 km^2. It caused channel filling and formed quake lake. Meanwhile, the mine was destroyed, the road was damaged, and the buildings collapsed.

Photo of landslides caused by the earthquake

◄ **Airborne optical remote sensing image of Chenjiaba Township, Beichuan County.**

This image was acquired on May 28, 2008. The image shows five large landslides along the river near Chenjiaba, marked A, B, C, D, and E. The landslides are 1.5 km^2, 0.54 km^2, 0.24 km^2, 0.2 km^2 and 0.15 km^2, respectively. Large area farmlands and forests were destroyed, many villages were buried, roads were damaged, and river courses were blocked by these landslides.

0 100 200 400m

▲ **Airborne optical remote sensing image of Guandili
Township, Pingwu County.**

This image was acquired on May 28, 2008. The image shows five landslides
(marked A, B, C, D, and E) on the hillside near Guandili, which destroyed many
farmlands. Landslide areas are 0.07 km², 0.98 km², 0.16 km², 0.05 km², and 0.06
km², respectively.

This image was acquired on May 28, 2008. A 0.3 km² landslide along the left bank near Daojiaoli destroyed farmlands.

▲　*IKONOS* optical remote sensing image near Wenchuan County

This image was acquired on May 23, 2008. The earthquake caused the massive collapse of the mountains near Wenchuan County. One of the largest collapse areas was 1 km^2 in the middle of the image. Vegetation on the mountain was destroyed.

◄　Airborne optical remote sensing image of Shibangou, Hongguang Township, Qingchuan County

This image was acquired on May 18, 2008. The hillside generated massive landslides near Shibangou, Hongguang Township. These landslides destroyed vegetation and brought clasts, which are very harmful. Under the action of rain erosion, clasts can collect into debris floods.

Photograph of earthquake debris flood.

► **Airborne optical remote sensing image of Dongjienao, Wenchuan County.**

This image, acquired on May 16, 2008, shows a high terrain slope location. Many landslides and collapses were generated in the earthquake. Collapse A destroyed the road. Collapse B destroyed a building. Landslide C threatened roads and the power station.

► **Airborne optical remote sensing image of Ganxipu, Yingxiu Township, Wenchuan County.**

This image was acquired on May 16, 2008. The upper right landslide, which is 0.2 km², formed a scarp that was about 20-m high. This landslide broke vegetation, damaged Minjiang River and roads, and even formed a debris-flow gully.

This image was acquired on May 23, 2008. The red color in the image indicates vegetation. A large-area landslide and collapse occurred in the mountain area of the riverbank of the Minjiang River near the town of Yingxiu. These geological disasters further induced mud-rock flow and debris flow. Vegetation was severely damaged. The road was broken into pieces in many places. Cropland and houses were also destroyed by the earthquake.

▼ Airborne optical image for the town of Jintiaogou, Maoxian County

This image, acquired on May 15, 2008, shows a large-area landslide and collapse over the mountain slope area on both sides of the major river courses, and it shows debris that flowed into the river valley. The landslide then formed a destructive mud-rock flow and debris flow as a result of rainfall in the river valley and in the tributaries of the major rivers. Crops and forests were extensively damaged. Parts of the houses collapsed. Roads were broken, and rivers were blocked.

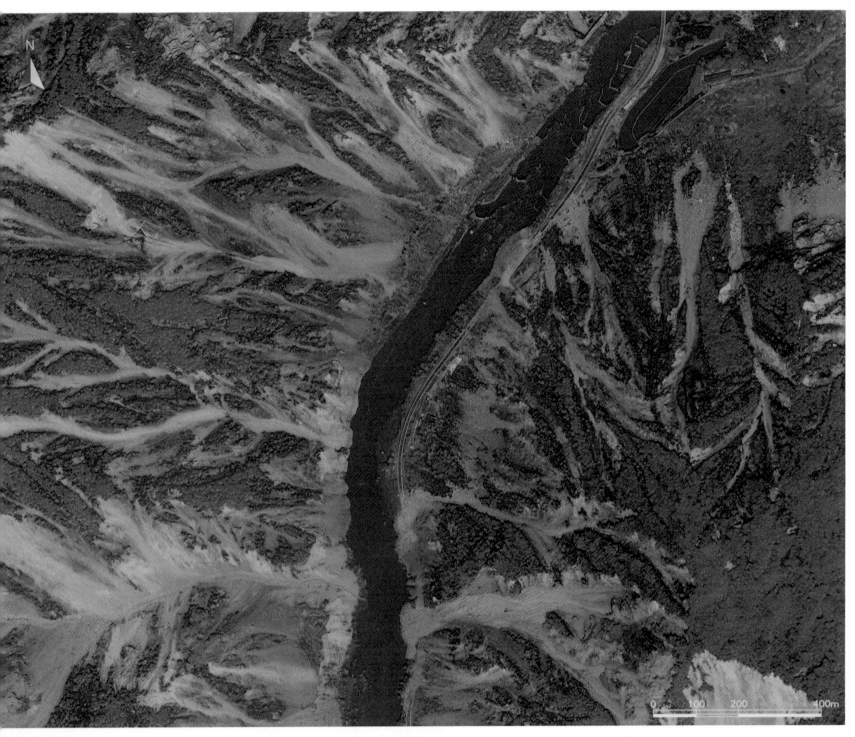

Photo of debris flow after the earthquake

▲ Airborne optical image of Shaba Village, Beichuan County

This image was acquired on May 16, 2008. Debris that formed along with the collapse flowed into the river valley and blocked the road. Vegetation was severely damaged, so that the river valley may further erode toward the upper river under the rainfall erosion effect.

◄ Airborne optical image of Bayi Village, Shiba township , Qingchuan County

This image, acquired on May 18, 2008, shows a series of fragment strips. Vegetation was extensively damaged surrounding the fragment strips. The collapse occurred over a large area, which creates a risk of mud-rock flows.

0   50   100   200m

▲   Airborne optical image of Xiaomaoping Village, Wenchuan County

The image was acquired on May 15, 2008. The image depicts the steep slope in the mountain area on the side of the Minjiang River. The Wenchuan Earthquake include mud-rock flows in the river valleys, which blocked the river and threatened the roads and the residential life.

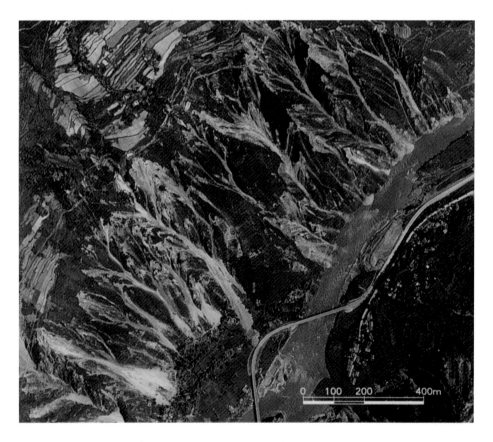

0   100   200   400m

◄   The airborne remote sensing optical imagery of Xindian, in Wenchuan County

From this image, captured on May 15, 2008, we can see that a large clastic flow gully was formed less than 50m away from the northwest bank of the Minjiang River.

This image was captured on May 15, 2008. There are four clastic flows marked A, B, C, and D. Areas of flows are 0.01, 0.015, 0.02 and 0.014 km^2, respectively. They washed off the road and destroyed the vegetation.

▼ **Airborne remote sensing optical imagery of Zesang in Wenchuan County**

This image was captured on May 15, 2008. Debris flows and clastic flows were formed on the west bank of the Minjiang River under the action of this earthquake. The vegetation was destroyed on both sides of the gully. The rainwater poses a particular threat in this landscape, and heavy rains threaten the road and riverway in the lower reaches of the river.

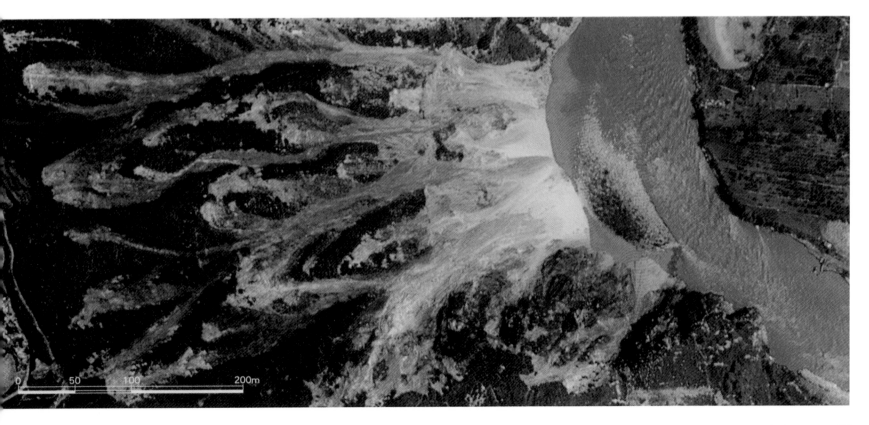

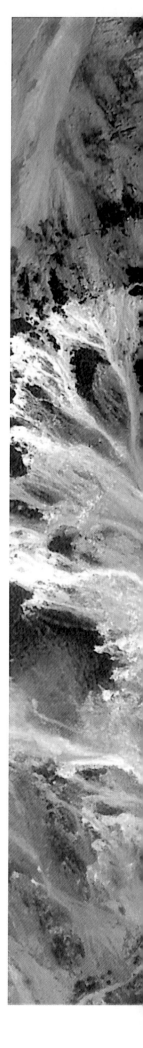

Photo of debris flow after the earthquake

▲ **Airborne remote sensing optical imagery of Maliuwan in Wenchuan County**

This image was captured on May 15, 2008. The large clastic flow gully was formed less than every 500 m on both banks of the Minjiang River, and vegetation was destroyed on both sides of the gully. The alluvial materials at the front of the gully blocked the riverway.

▶ **Airborne remote sensing optical imagery of Yangdian Village in Wenchuan County**

This image was captured on May 16, 2008. From the image we can see that the vegetation was destroyed on both banks of the Minjiang River and its branch, and the group clastic flows that formed along the side slope threaten the safety of the Minjiang River, the roads, and buildings. Many buildings were destroyed, and the roads were washed out. The riverways were narrowed by alluvial materials.

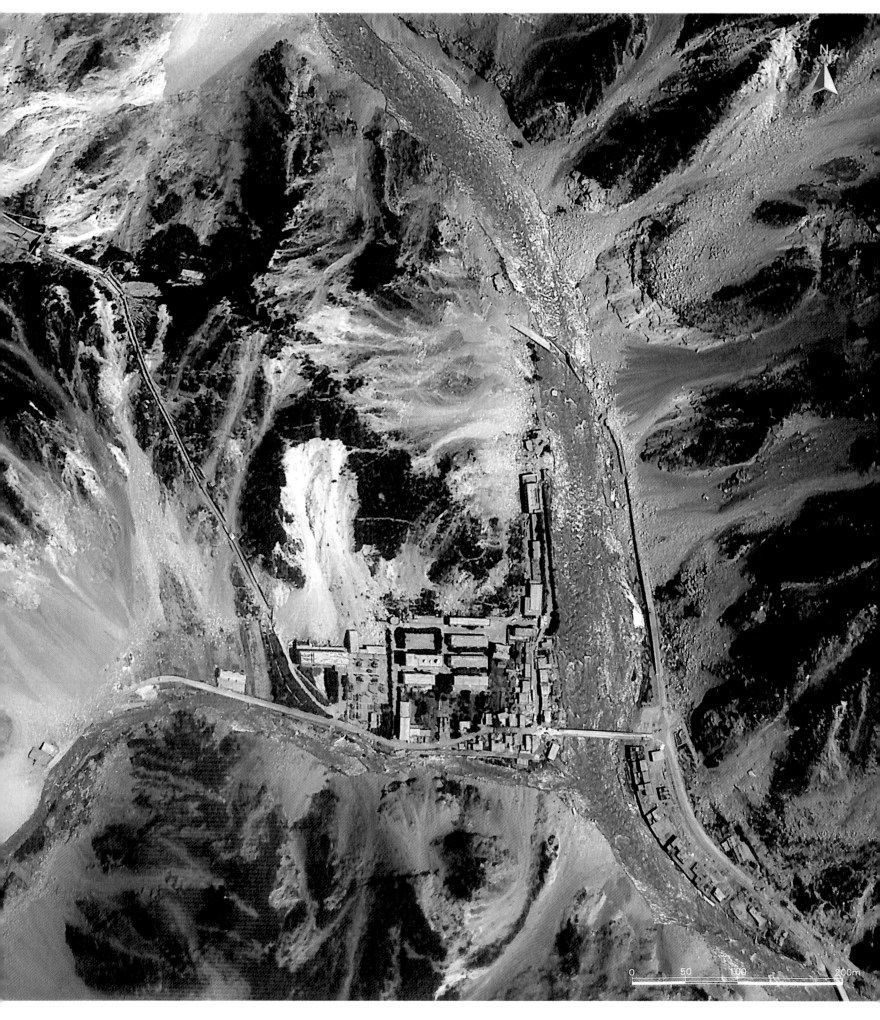

▲ Airborne remote sensing optical imagery of Gujing Village, in Qingchuan County

This image was captured on May 28, 2008. The large-area landslides were found near Gujing Village. Its area is 1.1 km². Large areas of farmlands were destroyed, the roads were buried, and the riverways were blocked by slump deposits. Then the barrier lake was formed.

◄ *ALOS* satellite imagery of Gaochuan Township in Anxian County

This image was captured on May 16, 2008. From the image we can see that the landslides were found near Gaochuan Township at the junction of Anxian County and Mianzhu City. The sliding mass marked A is located at 104°7'20"E, 31°38'37"N, its length is 4.89 km, and its area is 8.79 km². The sliding mass noted by B is located at 104°8'19"E, 31°33'10"N, and its length is 4.12 km, and its area is 4.78 km².

A photo of the town of Yingxiu after the earthquake

A building destroyed by debris flow

A building destroyed by the earthquake

Debris flow caused by the earthquake

Yingxiu Primary School after the earthquake

A building destroyed by the earthquake

Chapter

BARRIER LAKES

Barrier lakes are formed by water accumulation that occurs while a river valley or river bed is blocked by large-scale landslides. Barrier lakes are held in place by unstable dams, which could wash out, corrode, dissolve, or dissipate. The water level of a barrier lake can increase quickly, bringing more potential risks in the event of secondary disasters such as aftershocks and landslides. Barrier lakes can cause large-scale floods when dams collapse.

Many barrier lakes were formed by the Wenchuan earthquake, bringing huge risks to the areas downstream of the barrier lakes. Through analysis and evaluation of remotely sensed images, scientists can determine methods of harnessing barrier lakes. Optical images are intuitive and easy to understand, while radar images can be taken and viewed in all weather conditions. We have examined 46 barrier lakes in the disaster area. In this section we provide high-quality spaceborne and airborne remote sensing images of barrier lakes, including optical images with various resolutions and radar images with various resolutions and polarization conditions, such as spaceborne *TerraSAR-X, Radarsat-1 and -2, COSMO-SkyMed, IKONOS, SPOT 5*, etc., and airborne multi-polarization SAR and optical images with a spatial resolution of 0.5 m.

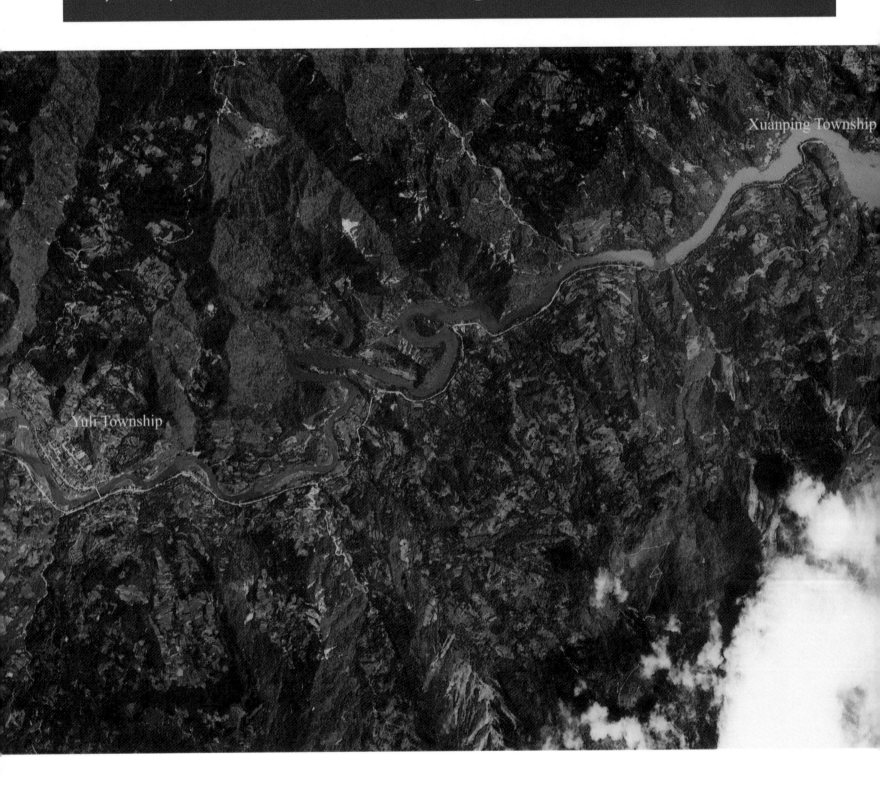

Xuanping Township

Yuli Township

Barrier Lake

Beichuan

▲   In this image, mud and rocks have fallen into the river because of a mountain landslide, and the channel is completely blocked. A large barrier lake formed in Tangjiashan, which is 3.2 km away from Beichuan County. Chaping Village, which lies upstream of the lake, was flooded, while Yuli Village was not flooded.

▲ Time series image of Tangjiashan in Beichuan County (acquired on May 16, 2008)

This is an airborne optical image of Tangjiashan in Beichuan County, acquired on May 16, 2008. In the image mud and rocks have fallen into the river as a result of a mountain landslide and debris flow. The channel in the south is completely blocked, and a large barrier lake has been formed in Tangjiashan. The elevation is 750.2 m on the top of the dam. The dam of the barrier lake itself is 82.8 m high and about 220 m long along the river. It is about 3,550 km² upstream of the lake. The dam is about 803 m long and 611 m wide at the widest place, and the area on the top is about 300,000 m².

▼ Time series image of Tangjiashan in Beichuan County (acquired on May 19, 2008)

This is an airborne optical image of Tangjiashan in Beichuan County, acquired on May 19, 2008. The water level is increasing because of the river supply. Until May 19, the water storage of Tangjiashan Barrier Lake was about 30 million m³.

▲ Time series image of Tangjiashan in Beichuan County (acquired on May 23, 2008)

This is an airborne optical image of Tangjiashan in Beichuan County, acquired on May 23, 2008. River input caused the water levels to increase constantly. As of 23 May, the water storage of Tangjiashan Barrier Lake was about 100 million m³. The water level increased 2.6 m compared to the day before.

▲ Time series image of Tangjiashan in Beichuan County (acquired on May 27, 2008)

This is an airborne optical image of Tangjiashan in Beichuan County acquired on May 27, 2008. River input caused the water levels to increase constantly. At 8 am on May 27, the elevation of the water level in Tangjiashan Barrier Lake was 727.02 m, an increase of 1.82 m compared to the water level at 8 am on May 26, 2008. It was 23.18 m from the water surface to the top of the dam of the barrier lake. The water depth near the dam is 60.37 m. In the image, disaster-relief tents are clear to the north of the dam. Large equipment is seen working in the diversion channel excavating the field. The diversion channel was excavated 180 m at that time.

May 24, 2008

May 25, 2008

Jianjiang River

Time series of airborne radar images of Tangjiashan in Beichuan County

This is a series of airborne radar images of Tangjiashan in Beichuan County, acquired on May 24, 25, 26 and 31, 2008. In the time series images, the channel is completely blocked by the barrier lake dam as a result of a large mountain landslide. There are small changes in the volume of the dam locally, mainly in the eastern part of the dam. But overall, the changes are negligible, and it remains a barrier lake with high risk. In the image acquired on May 31, the excavating diversion channel is partially complete, and the project in the west is nearly complete. At this time there are 6 or 7 m between the upper water level and the bottom of the diversion channel.

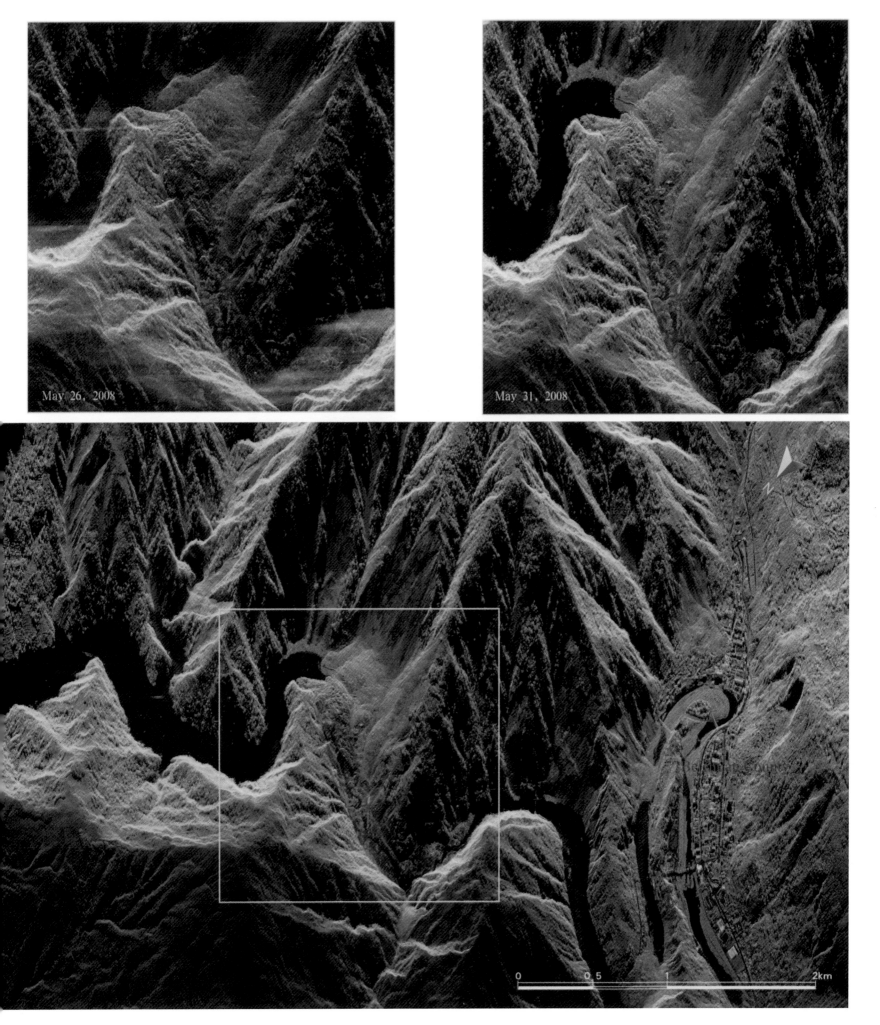

May 26, 2008

May 31, 2008

0 0.5 1 2km

Above is a R*adarsat-1* image acquired on December 19, 2007 (before the earthquake) around the Tangjiashan Barrier Lake in Beichuan County. Below is a *Radarsat-1* image acquired on May 14, 2008 (after the earthquake) around the Tangjiashan Barrier Lake in Beichuan County. The changes are clear from the comparison of the two images: barrier lakes and cofferdams are formed, blocking the river channels in Jianjiang Basin because of landslides, debris flow, and some other factors.

Radarsat-1 image acquired on May 14, 2008

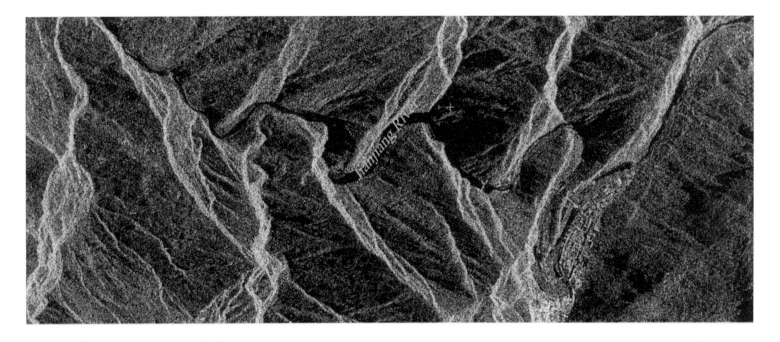

Radarsat-1 image acquired on December 19, 2007

Jianjiang River

Beichuan County

SPOT 5 image acquired on November 10, 2006

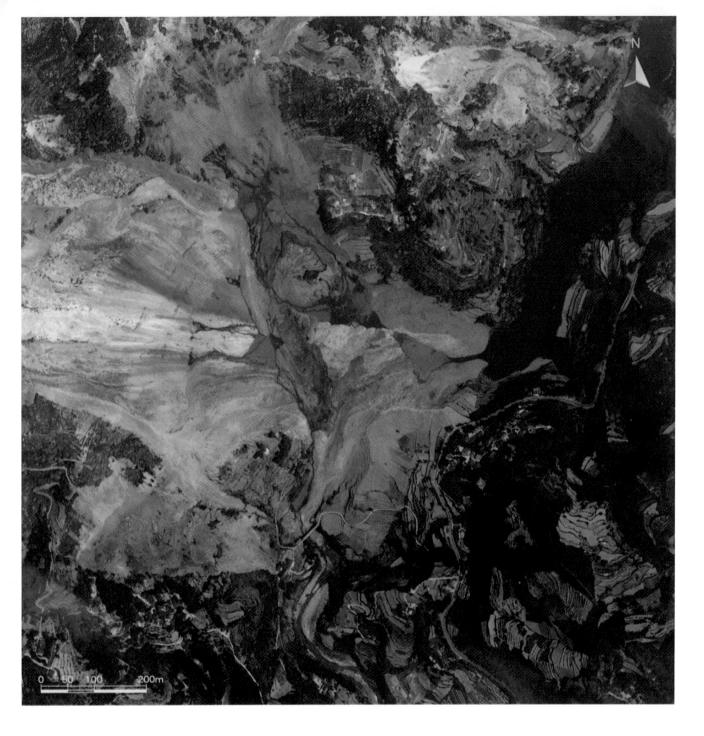

▲ Airborne optical image of a barrier lake in Xiaojiaqiao

▶ Airborne optical image of the barrier lake in Donghekou, in Qingchuan County

This is an airborne optical image of Xiaojiaqiao, acquired on May 19, 2008. Large barrier lakes formed in Xiaojiaqiao because of landslides during the earthquake. The river channel was completely blocked by the dam, and this barrier lake is one of the most dangerous lakes. The water volume of the lake is more than 10 million m^3, and the water level near the dam is more than 50 m. The dam formed by mountain landslides is 198 m wide and 200 m long.

This is an airborne optical image of Donghekou in Qingchuan County, acquired on May 28, 2008. The image shows that many mountain landslides occurred in the area of Donghekou, in Hongguang Village, in Qingchuan County as a result of the earthquake, and the landslide volume reached more than 20 million m^3. Channels are blocked, and three barrier lakes, Donghekou, Hongshi River, and Shibangou, with a total volume of 12 million m^3, were formed in the Qingzhu and Hongshi Rivers. Although the Donghekou Barrier Lake is not too large, the river is completely blocked.

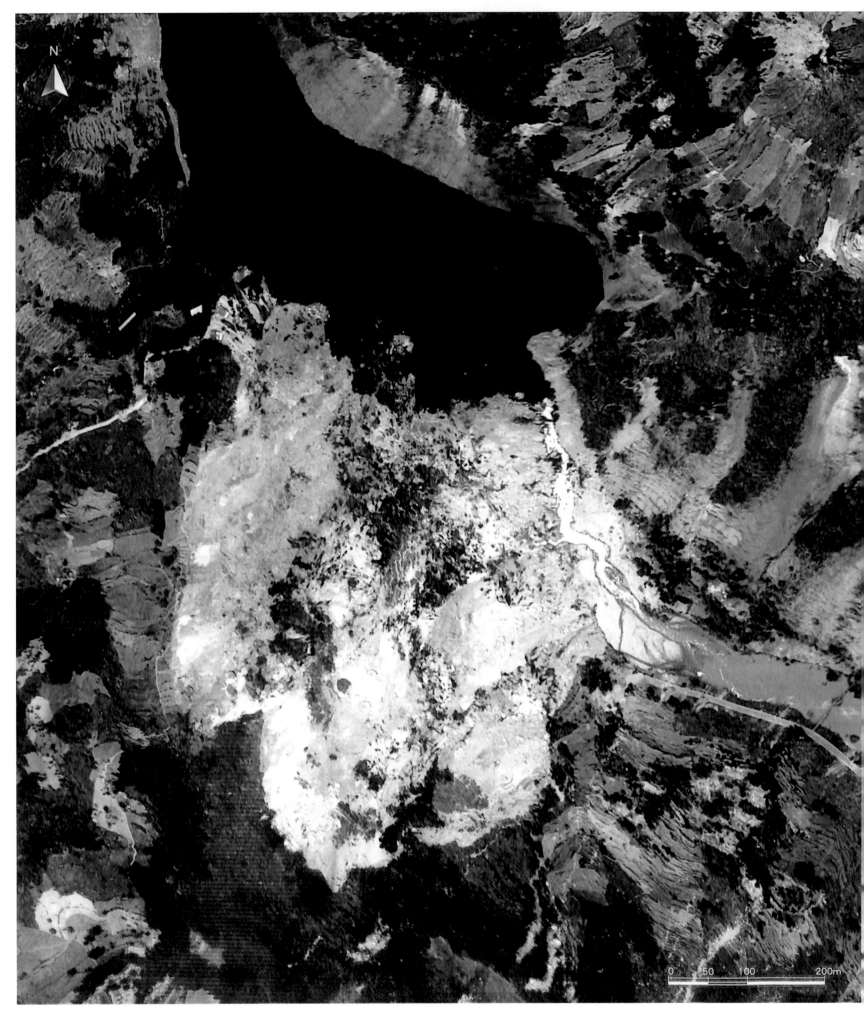

▼ Airborne optical image of the barrier lake in Kuzhuba, in Beichuan County

◄ Airborne optical image of a barrier lake in Shibangou, in Qingchuan County

This is an airborne optical image downstream of Kuzhuba in Beichuan County. Barrier lakes formed because of large mountain landslides blocking the river channel. The massive barrier body is heaped south of the channel and completely blocks the channel. The dam is stable because it is formed of hard stones, so dredging the river channel is very difficult.

This is an airborne optical image of Shibangou in Qingchuan County and Hongshi River, acquired on May 28, 2008. The large barrier lake is formed because of a large mountain landslide blocking the river channels. The lake lies 3 km away from Donghekou Barrier Lake upstream of Qingjiang River with a volume of 8 million m³, and it is formed by 10 million m³ of mountain landslide in both sides. The dam is 75 m high, 450 m wide, with a drop of 60 m.

▶ Airborne optical image of the barrier lake of Yanyangtan, in Beichuan County

This is an airborne optical image of Yanyangtan in Beichuan County, acquired on May 28, 2008. The image shows that large mountain landslides moved into the river channel. On the east side the river still flows, although there is a small hill mass in the river. On the west side the channel is directly blocked by a large mountain body, as a result, a clear barrier lake is formed.

▶ Airborne optical image of the barrier lake in Guanzipu, in Qingchuan County

This is an airborne optical image of Guanzipu in Beichuan County, acquired on May 28, 2008. The barrier lake is formed south of the river as a result of large mountain landslides, debris flow, and other geological disasters that blocked the river channel. The volume of this lake is relatively small, and its water continues to flow in some places. The part of the river channel that is completely blocked is small.

A. Jianjiang River

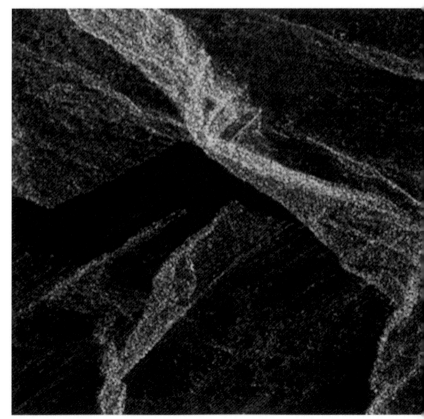

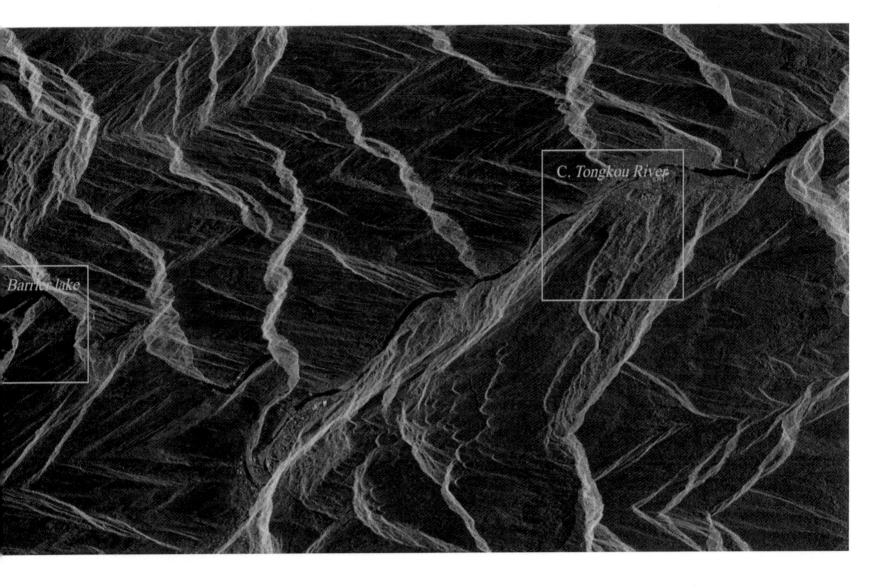

Barrier lake

C. Tongkou River

A *TerraSAR-X* satellite based radar image of Jianjiang and Tongkou River Basin

This is the German *TerraSAR-X* satellite-based radar image of the Jianjiang and Tongkou River basins acquired on May 17, 2008. It is clear that three barrier lakes formed as a result of large mountain landslides rushing into the river channel. They lie in (A) Jianjiang, (B) Beichuan, and (C) beside Tongkou River. The largest one is Beichuan Barrier Lake, the second is Tongkou River, and the third one is Jianjiang Barrier Lake.

Barrier lake

Tianchi Village

▲ Airborne SAR image of the village of Tianchi, in the town of North Hanwang

This is an airborne SAR image of the village of Tianchi, in the town of Hanwang, in the city of Mianzhu, acquired on May 24, 2008. The image shows several barrier lakes that formed along the Jinyuan River as a result of mountain landslides blocking the river channel. More than 20 km away from Hanwang Town in the north, two mountains shut off the river and a 5-km-long barrier lake formed.

This is an airborne optical image of Changheba in Mianzhu City acquired on May 23, 2008. In the image, a barrier lake that formed as a result of large mountain landslides blocking the river channel is shown. The river channel is completely blocked. It is clear that the barrier lake formed in the east part of the river in the image.

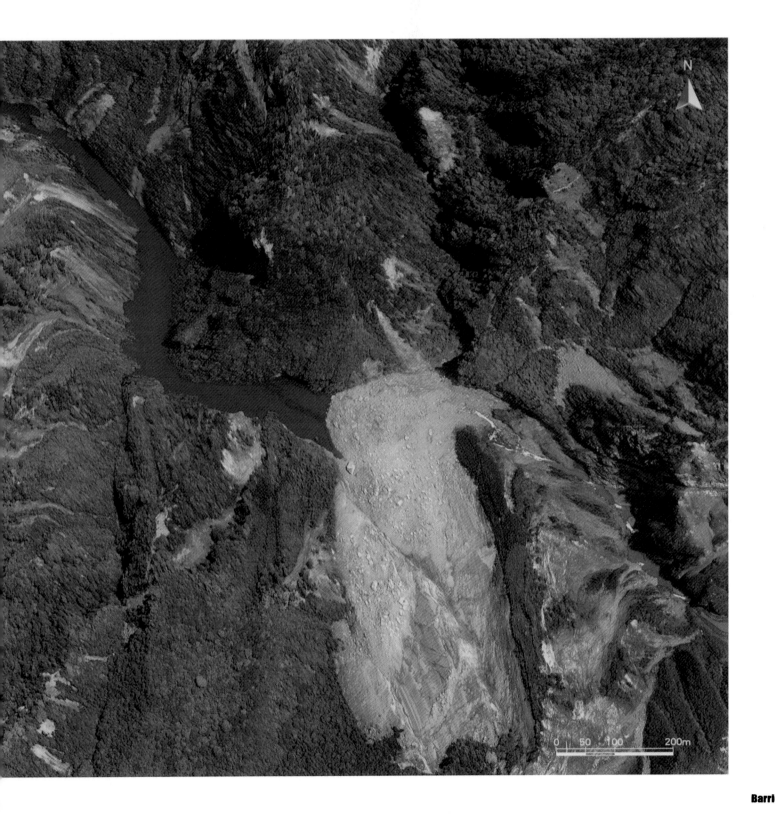

▶ **Airborne SAR image of Ganhekou and Machaotan in Shifang City**

This is an airborne SAR image of Ganhekou and Macaotan in the city of Shifang acquired on May 12, 2008. In the image, several collapses occurred on both sides of the upper Shiting River. Four barrier lakes, marked A, B, C, and D, formed in Shihekou and Macao Beach. The two channels in C and D have been completely blocked, while channels in A and B become noticeably narrow as a result of the blocking barrier body.

This is an airborne optical image in Pingwu County acquired on May 28, 2008. In the image, several mountain landslides and debris flows occured in Wenjiaba, Pingwu County, in Sichuan province, blocking the river channel, and then a barrier lake formed. The river in the northeast still flows, although a relatively small hill mass lies there. The channel in the southwest is completely blocked by the landslide body.

This is an airborne optical image at Wenjiaba in Pingwu County acquired on May 28, 2008. In the image, Wenjiaba lies in the branch of Fujiang, Pingwu County, and the river channel is blocked by mountain landslides as a result of the earthquake, so a large barrier lake formed. Before flood discharge, the lake was 100 m wide.

Barrier Lake

0   150   300   600m

▲ Satellite radar image *COSMO-SkyMed* in Tongkou River of Beichuan County

This radar image is from an Italian satellite, *COSMO-SkyMed*, with X band and HH polarization in the Tongkou River Basin, acquired on May 14, 2008. The image shows several barrier lakes that formed as a result of large landslides blocking the river channel, in which the one besides Beichuan County is the largest and most seriously blocked.

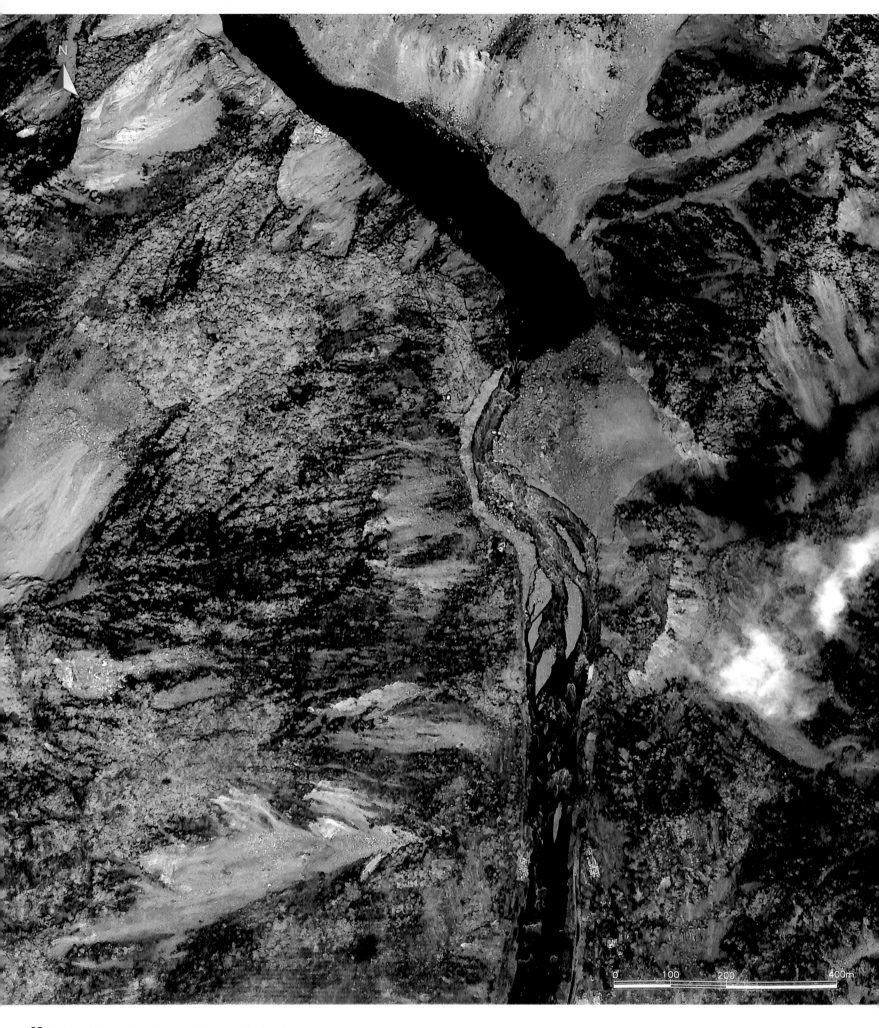

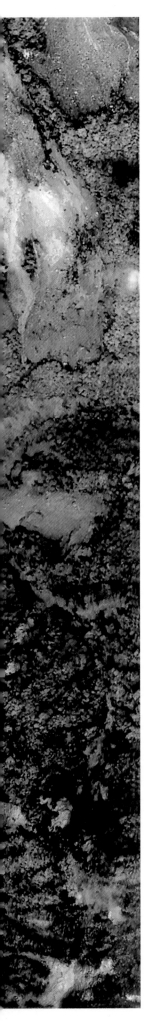

▲ This photo shows construction sites that were submerged because of water levels rising and the Minjiang River becoming narrower because of landslides and collapses.

Chapter

COLLAPSED BUILDINGS

The Wenchuan earthquake and the secondary disasters induced by the earthquake caused massive collapses of buildings, enormous losses of property, and tens of thousands of casualties. This chapter looks at collapsed houses and buildings based on airborne optical images. Airborne radar and optical remote sensing images of the region most severely affected by the earthquake were acquired during May 15-May 28. These images cover 14 counties, including Wenchuan, Beichuan, Mianyang, Shifang, Qingchuan, Maoxian, Anxian, Dujiangyan, Pingwu, Pengzhou, Lixian, Jiangyou, and Guangyuan. To estimate the rates of building collapse, high-spatial satellite data from sources such as *IKONOS*, *QuickBird*, *WorldView*, and *SPOT 5* before and after the earthquake were studied, and high-spatial airborne SAR images were used for interpretation.

Using high-resolution airborne images, we studied collapsed buildings in 102 villages and towns in 14 counties and acquired the collapse ratio contour maps of the 14 counties. We compiled these data with the earthquake fault zone data of Longmen Mountain to determine the typical patterns of building collapse. The house collapse ratio in the disaster region was monitored and assessed visually.

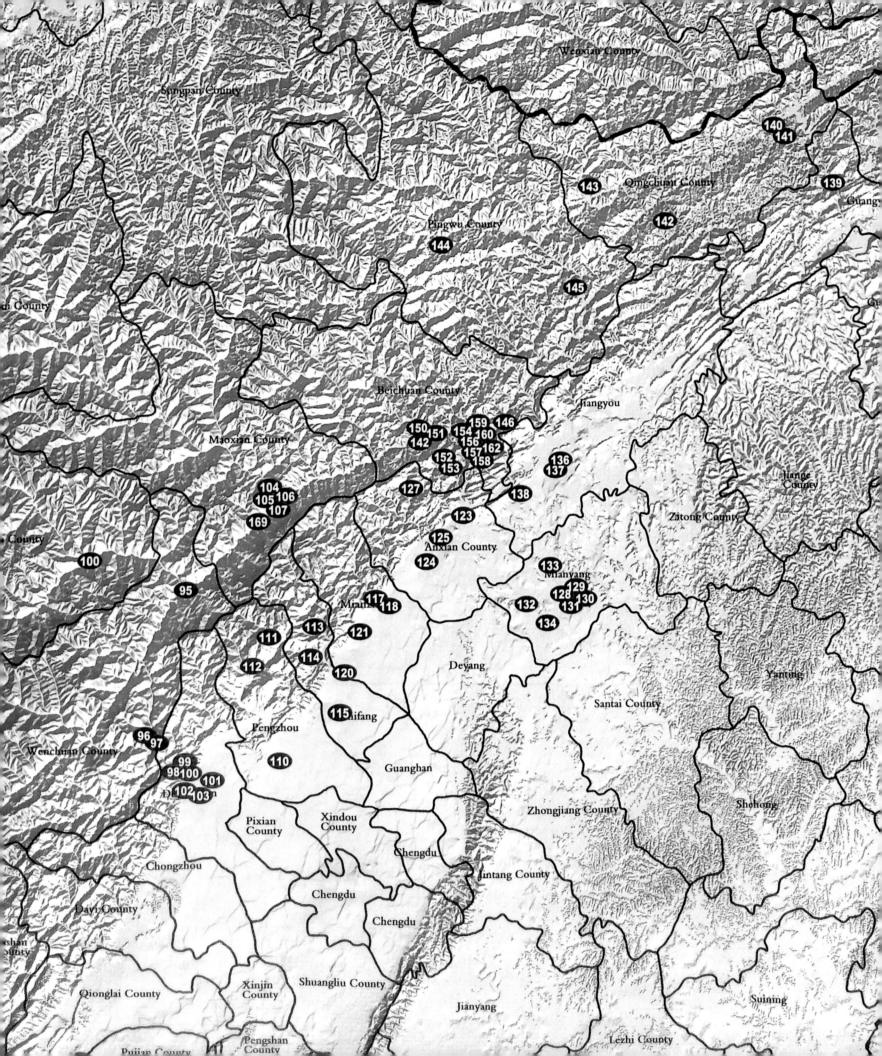

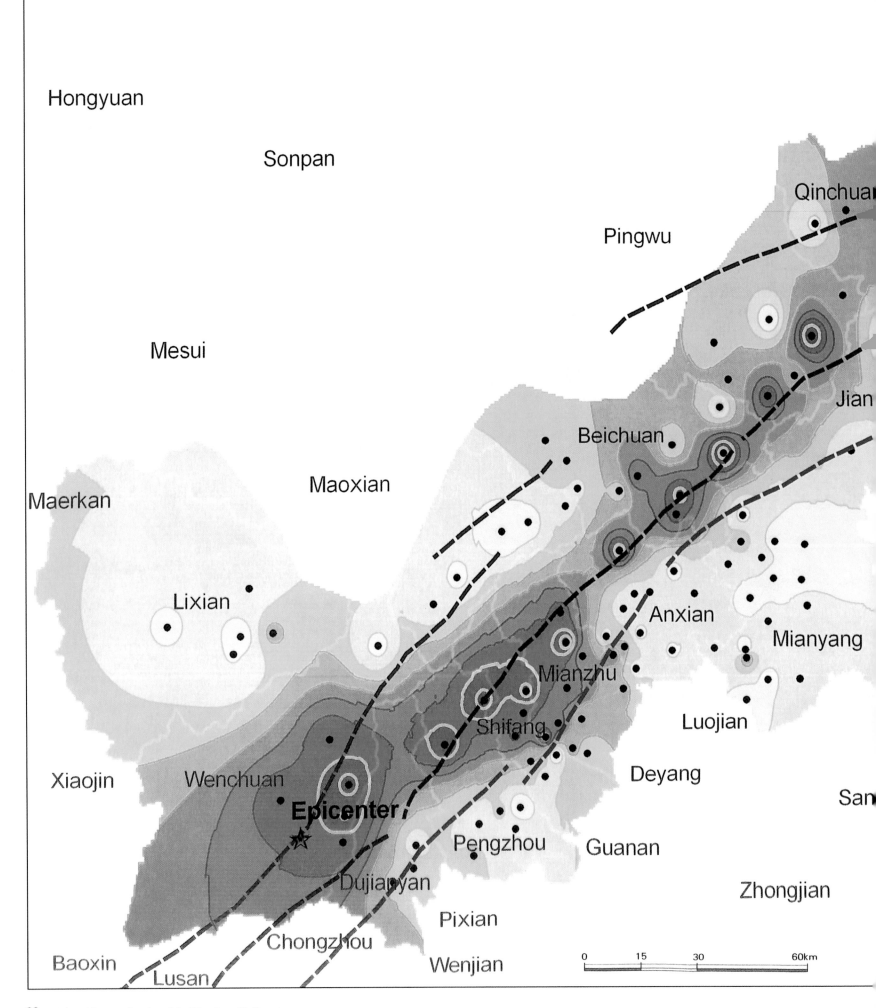

Hongyuan

Sonpan

Qinchua

Pingwu

Mesui

Jian

Beichuan

Maoxian

Maerkan

Lixian

Anxian

Mianyang

Mianzhu

Luojian

Shifang

Xiaojin   Wenchuan

Deyang

Epicenter

San

Pengzhou   Guanan

Dujianyan

Zhongjian

Pixian

Baoxin

Chongzhou   Wenjian

Lusan

0   15   30   60km

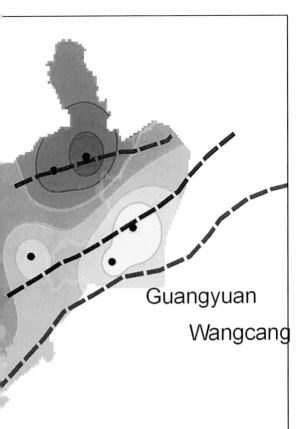

Guangyuan

Wangcang

Jiange

Cangxi

Langzhong

Collapsed houses in the main quake-stricken region

Using high-resolution airborne images, we studied collapsed buildings in 102 villages and towns in 14 counties and cities and created a contour map showing the collapse ratio in those counties and cities. From this contour map, together with the Longmenshan Fault data, we can clearly see the collapse pattern for houses and buildings.

(1) The most seriously hit villages and towns are located in the Longmenshan Fault, and they are distributed along the hill-front fault (Anxian-Guanxian Fault) and hill-back fault (Wenchuan-Maoxian Fault) within the fault zone.

(2) On the whole, the most seriously affected villages and towns are distributed along the central fault of the Longmenshan Fault, namely, the Beichuan-Yingxiu Fault.

(3) A small quantity of villages and towns suffering severe building collapses are located on the northwest side of the Longmenshan Main Fault, while the majority of them are on the southeast side of the main fault.

(4) In Pingwu County, the city of Guangyuan, and Qingchuan County, areas of severely damaged houses shifted from the Yingxiu-Beichuan Fault to the Wenchuan-Maoxian Fault. On the same cross-section, the amount of destruction is much worse in the town of Shazhou, in Qingchuan County (located on the Wenchuan-Maoxian Fault) than that in the town of Shadui (located on the Yingxiu-Beichuan Fault).

▲ Earthquake-induced clastic flow

▶ **Distribution of collapsed houses in an urban part of Wenchuan County**

Wenchuan County is not only the home of the Qiang people, well known for their traditional needlework, but also the home of the giant panda, which is considered to be a living fossil. Wolong, the most prestigious research center for giant pandas in the world, is located in the southwestern part of Wenchuan County.

Only 46 km away from the town of Yingxiu, which was the epicenter of the Wenchuan earthquake and located on the Yingxiu-Maoxian Fault, Wenchuan County was seriously damaged in the earthquake. Approximately one-sixth of all the houses in urban areas of the county were ruined.

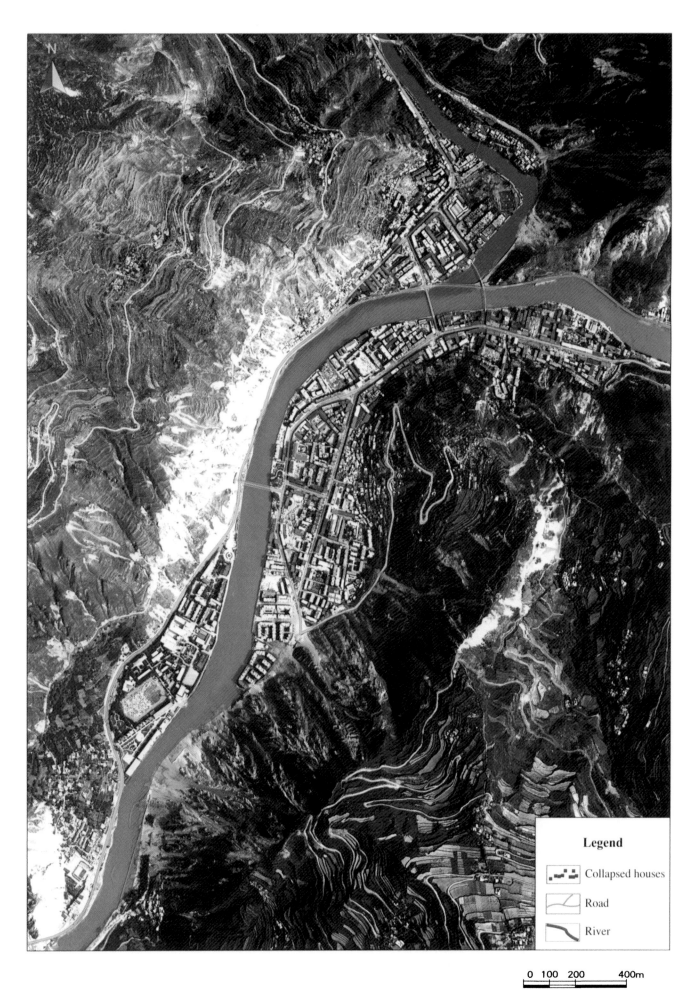

Legend

Collapsed houses

Road

River

0 100 200 400m

◀ Xuankou High School after the earthquake

◀ The town of Xuankou after the earthquake

◀ Airborne optical images of the town of Yingxiu, Wenchuan County

Acquired May 16, 2008. Located in the Yingxiu-Maoxian Fault, the town of Yingxiu was the epicenter of the Wenchuan earthquake. It was seriously damaged in the earthquake, over 90% of the houses in the town of Yingxiu collapsed. The red-frame in the image marks the severe damaged Xuankou High School.

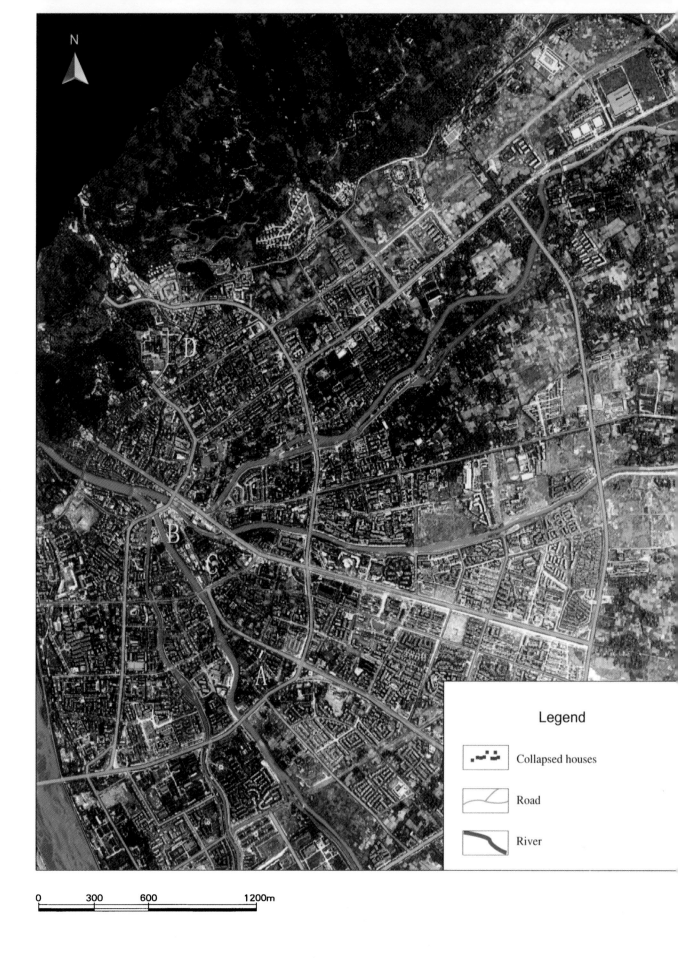

Legend

⬚ Collapsed houses

⬚ Road

⬚ River

0   300   600   1200m

Post-earthquake **Pre-earthquake**

Disaster assessment on building collapse in the city of Dujiangyan

The city of Dujiangyan lies on the northwest edge of the Chengdu Plain. It is 48 km away from the center of the city of Chengdu, and about 20 km away from the epicenter the town of Yingxiu. Pre-disaster *QuickBird* satellite images and post-disaster airborne images showed serious earthquake damage. About 10% of the houses in downtown Dujiangyan collapsed.

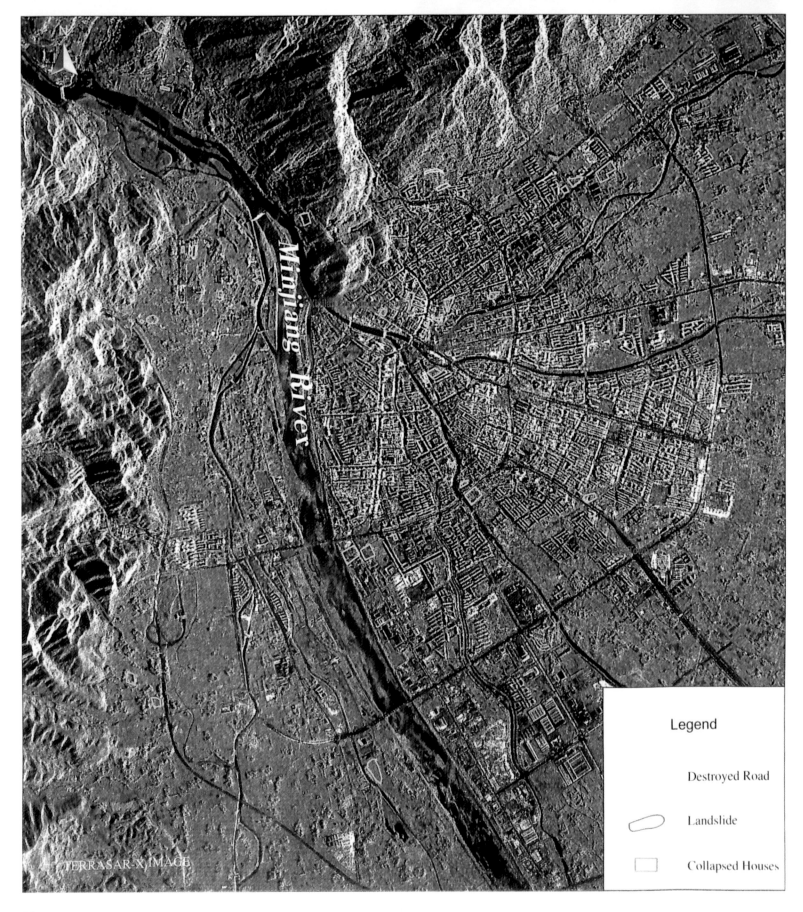

Minjiang River

TERRASAR-X IMAGE

Legend

Destroyed Road

Landslide

Collapsed Houses

Distribution of the collapsed buildings in Maoxian County

Maoxian County is located in the southeast section of Aba Tibet and Qiang Autonomous Prefecture, which is the transitional zone from the Qinghai-Tibet Plateau to the Sichuan Plain, and is the main residential area of the Qiang people. Maoxian County lies along the Wenchuan-Maoxian seismic belt, and nearly 20% of the houses were either severely damaged or collapsed in the urban area of the Maoxian County.

Legend

Collapsed houses

Road

River

0 0.125 0.25   0 5m

▲ Airborne optical image of the town of Longfeng, in the city of Pengzhou

Acquired May 19, 2008. The town of Longfeng, located in the piedmont plain of Longmen Mountain, is famous for producing garlic. It is 6 km away from the Anxian-Guanxian Fault and 19 km away from the Yingxiu-Beichuan Fault. About 20% of the houses in the town of Longfeng collapsed.

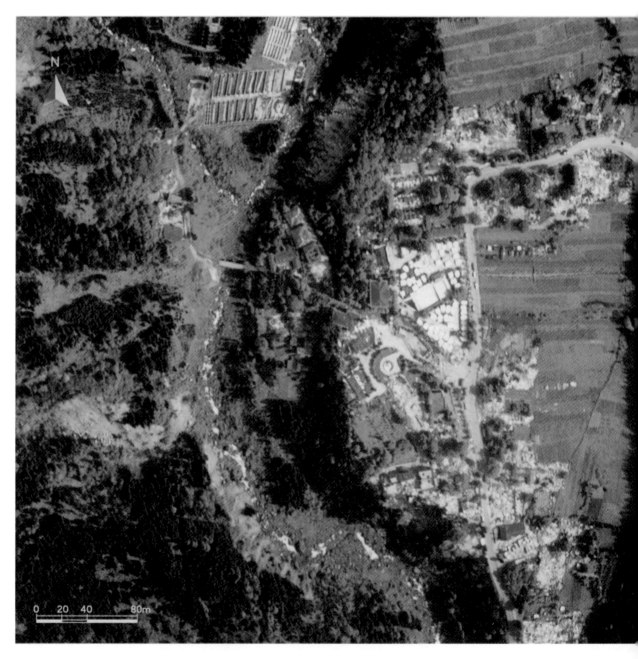

0   200   400   800m

◄ **Airborne optical imag**
of the town of Guangji,
the city of Mianzhu

Acquired May 19, 2008. The town
Guangji is in the piedmont transiti
region of Longmen Mountain a
between the Yingxiu-Beichuan Fa
and the Anxian-Guanxian Fau
Approximately 75% of the houses
the town of Guangji collapsed. The r
frame shows collapsed and seriou
damaged houses.

0 100 200 400m

▲ Collapsed building in the

town of Guangji

► Airborne optical image of the town of Zundao, in the city of Mianzhu

Acquired May 19, 2008. The town of Zundao is in the piedmont transition region of Longmen
Mountain, and between the Yingxiu-Beichuan Fault and the Anxian-Guanxian Fault. The
houses in the town of Zundao were seriously damaged, and the collapse rate approached
75%. The red frame shows the collapsed and damaged houses.

Distribution of collapsed buildings in Anxian County

Anxian County is located in the northwest portion of the Sichuan Basin, with a population of 500,000 and about 1,404 km² of land. Anxian County was seriously damaged in this earthquake. Analysis based on airborne images shows that about one-seventh houses in Anxian County collapsed; houses along both the Yingxiu-Beichuan Fault and the Anxian-Guanxian Fault were destroyed. The towns of Gaochuan, Qianfo, Chaping, Xiaoba and several other towns that lie across the Yingxiu-Beichuan Fault suffered 60% collapse rate.

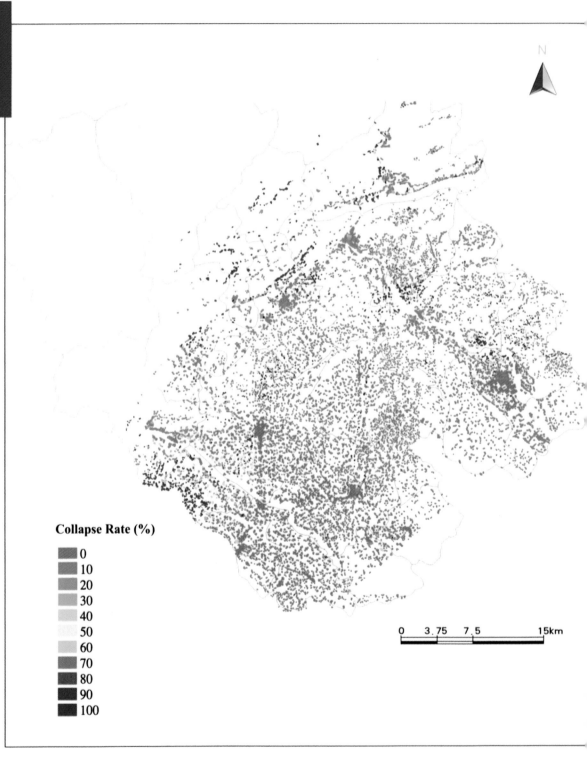

Collapse Rate (%)

- 0
- 10
- 20
- 30
- 40
- 50
- 60
- 70
- 80
- 90
- 100

0   3.75   7.5   15km

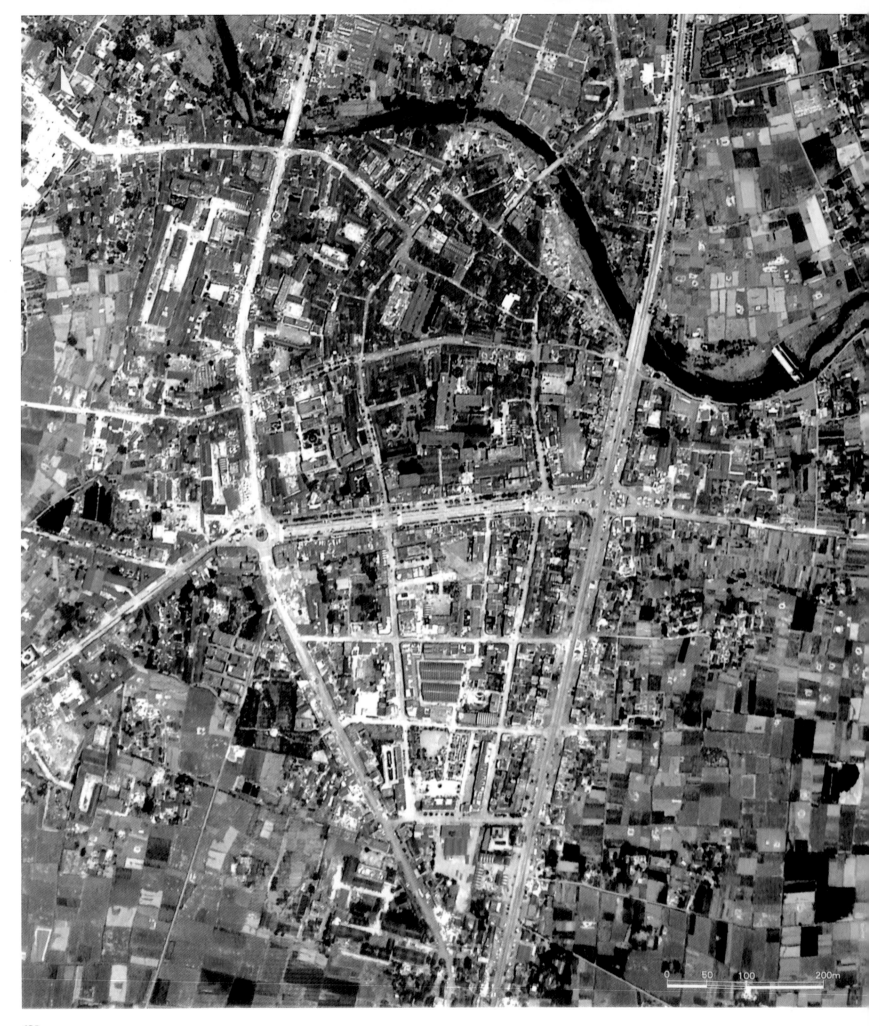

0 40 80   160m

Acquired May 18, 2008. The distance from Xinping to the urban area of Beichuan County is just 20 km. The houses in the town of Xinping were severely damaged, and the house collapse rate was over 25%.

Acquired May 18, 2008. The town of Sandui lies on the Yingxiu-Beichuan Fault but only a few houses, less than 10%, collapsed.

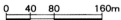

0   40   80   160m

▲ Airborne optical image of the town of Chenjiaba, in Beichuan County

Acquired May 28, 2008. In Chenjiaba, a town crossed by the Yingxiu-Beichuan Fault, the houses were heavily damaged, with over 80% of them destroyed. (A) A landslide split the town into two parts and caused heavy casualties. (B) Collapsed houses.

▶ *IKONOS* image of the town of Yuli, Beichuan County

Acquired May 23, 2008. The image shows that the floodwater of the Tangjiashan Barrier Lake began to submerge the town of Yuli.

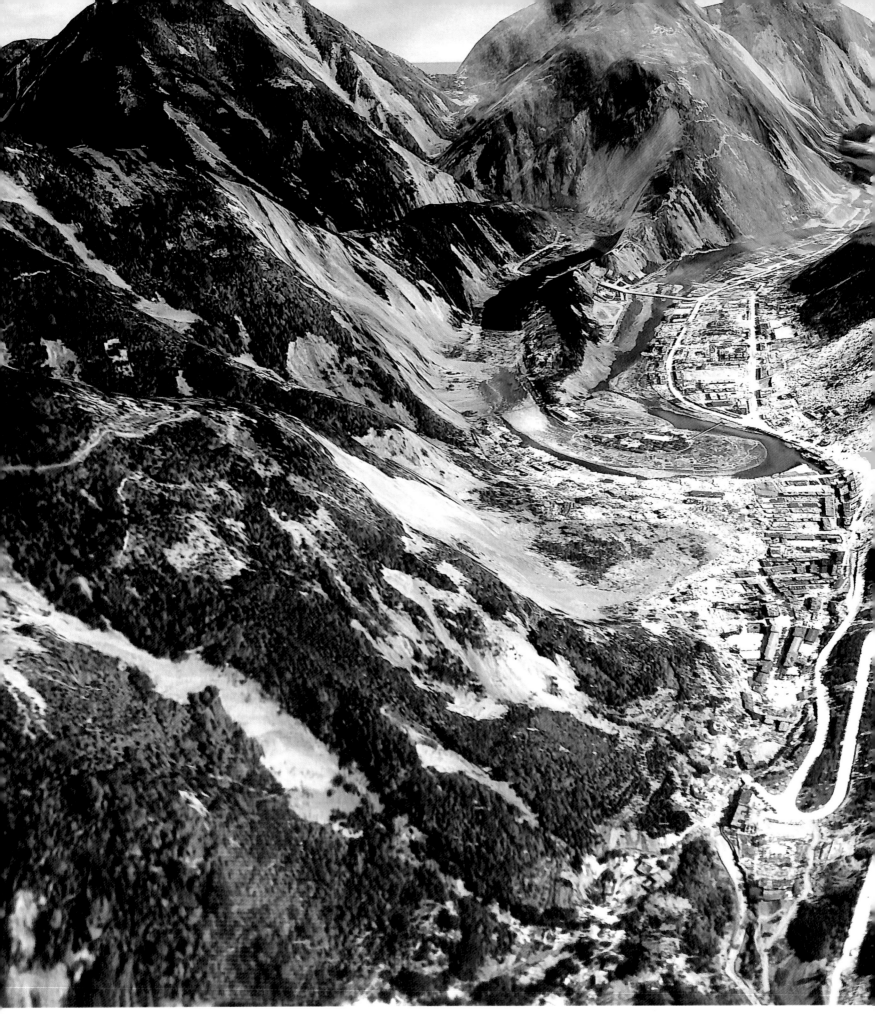

P. Road 302

Minjiang River

A

B

C

0   75   150   300m

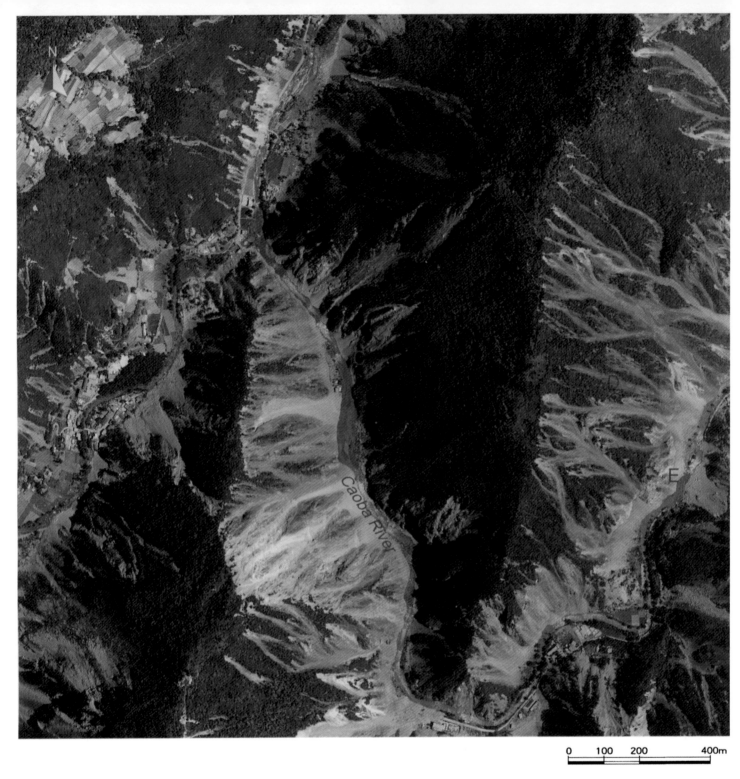

N. Road 213

Minjiang River

D

C

E

A

B

0   75   150   300m

Minjiang River

E

A

B

G.N. Road 213

D

Minjiang River

N. Road 213

0 40 80 160m

▲ Airborne ADS40 image of the village of Wenzhen, in the town of Nanxin, in Maoxian County

In the middle of this ADS40 image acquired on May 15, 2008, there was a chain of landslides along the east Minjiang River bank, which then caused destruction to National Road 213 for over 1,000 m. As shown on the image, (A), (B), and (C) marked the damaged sections, with the longest section (A) up to 230 m. The extent of the road blockage made quick repair impossible, as a result people were traversing the roadway by foot, but cars could not get through.

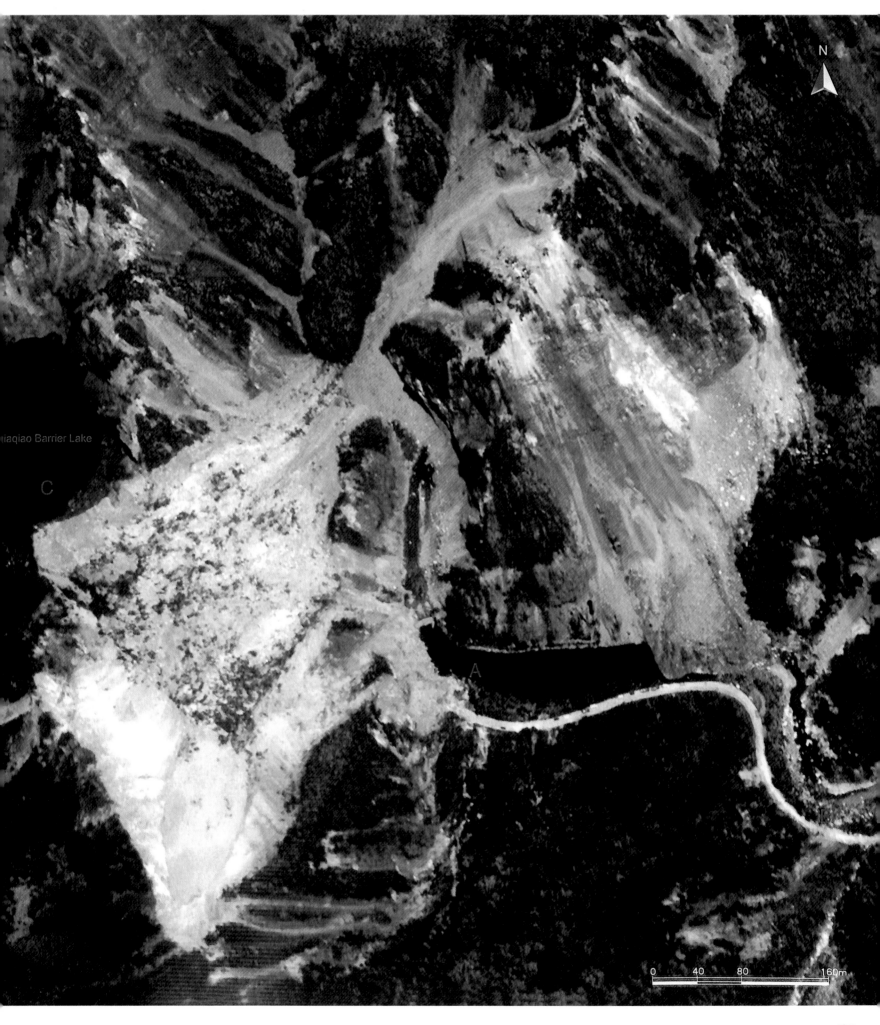

▶ Photo of xiaoyudong Bridge after the earthquake

This image shows the region from the town of Xiaoyudong to the town of Longmenshan that was affected by the earthquake and debris flow. Several parts of roads and railways through this region were ruined. (A) A road blocked by debris flow. (C) The collapsed Xiaoyudong Bridge (see the in situ photo), which lies in the northwest section of the city of Pengzhou.

Jianjiang River

0 125 250   500m

317 N. Road

213 N. Road

Minjiang River

317 N. Road

N

0 100 200   400m

▼ Airborne radar image over the urban district of Wenchuan
County (May 15, 2008)

This image shows national highways G213 and G317 passing through Wenchuan County blocked
by mountain landslides. (A) A damaged segment of G317, which leads to Lixian County. (B) Part
of the highway that was blocked. (C) A portion of G213 that connects Wenchuan County with the
city of Dujiangyan that was completely blocked by a huge landslide.

Road damage in Qingchuan County

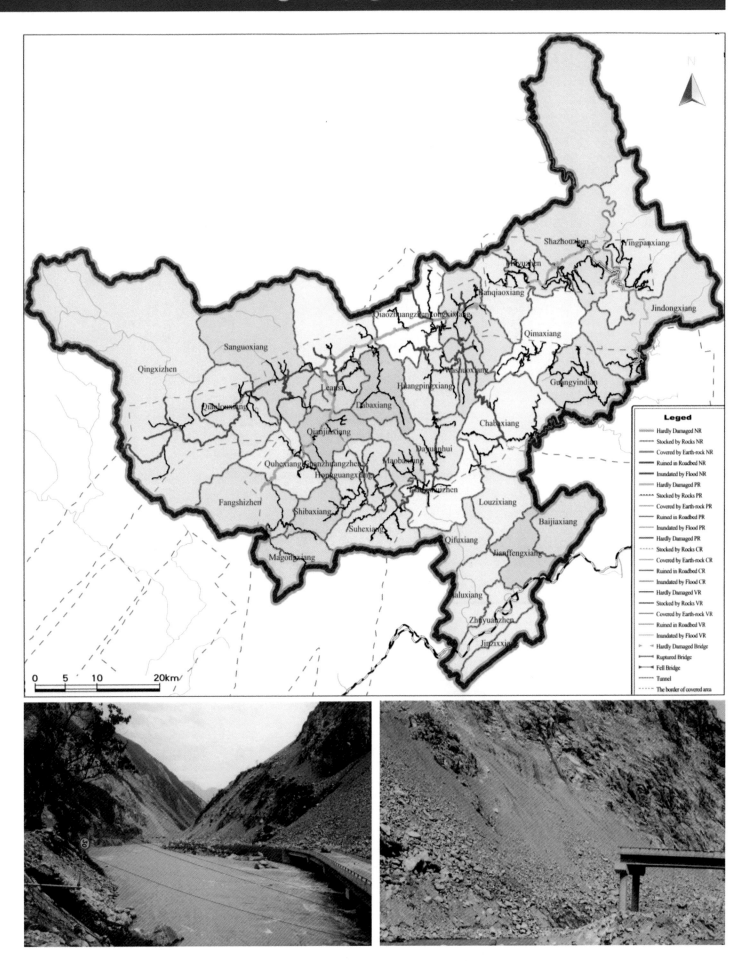

Leged

Hardly Damaged NR
Stocked by Rocks NR
Covered by Earth-rock NR
Ruined in Roadbed NR
Inundated by Flood NR
Hardly Damaged PR
Stocked by Rocks PR
Covered by Earth-rock PR
Ruined in Roadbed PR
Inundated by Flood PR
Hardly Damaged PR
Stocked by Rocks CR
Covered by Earth-rock CR
Ruined in Roadbed CR
Inundated by Flood CR
Hardly Damaged VR
Stocked by Rocks VR
Covered by Earth-rock VR
Ruined in Roadbed VR
Inundated by Flood VR
Hardly Damaged Bridge
Ruptured Bridge
Fell Bridge
Tunnel
The border of covered area

0   5   10   20km

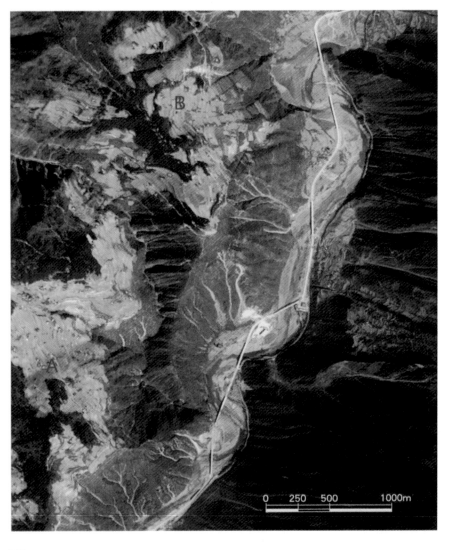

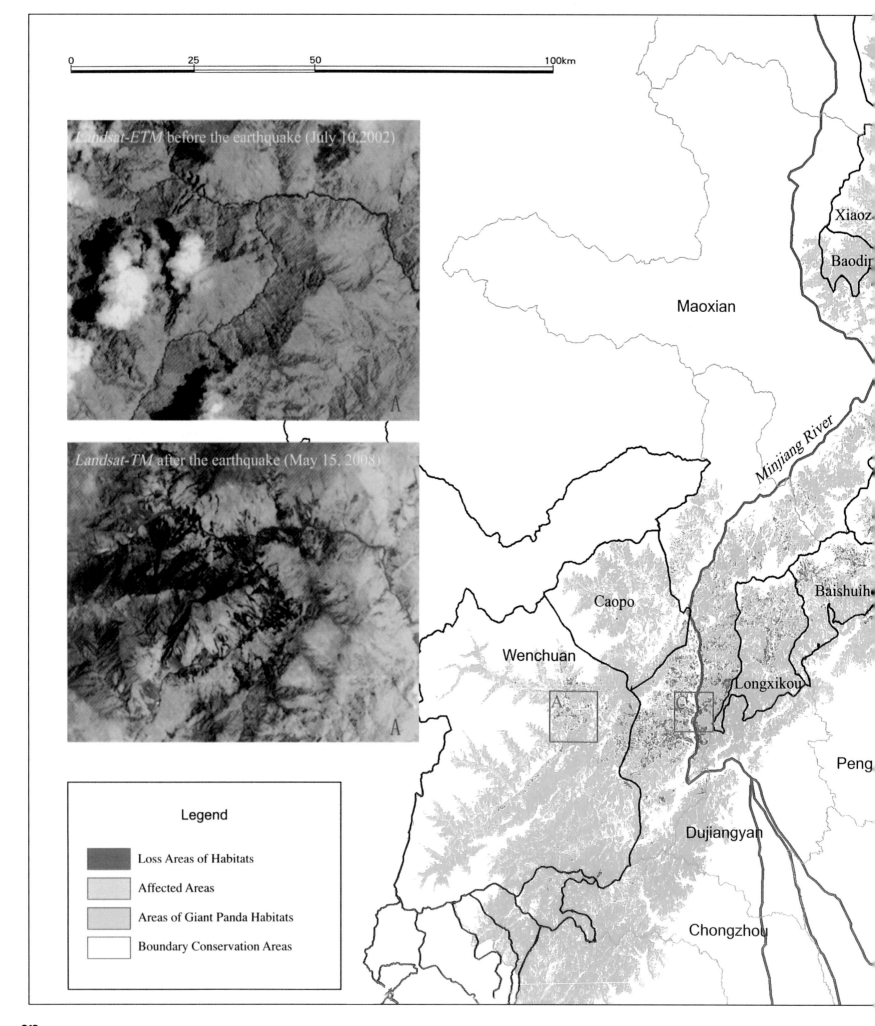

Landsat-ETM before the earthquake (July 10,2002)

Landsat-TM after the earthquake (May 15, 2008)

Xiaoz

Baodir

Maoxian

Minjiang River

Baishuih

Caopo

Wenchuan

Longxikou

Peng

A

C

Dujiangyan

Chongzhou

Legend

Loss Areas of Habitats

Affected Areas

Areas of Giant Panda Habitats

Boundary Conservation Areas

0   25   50   100km

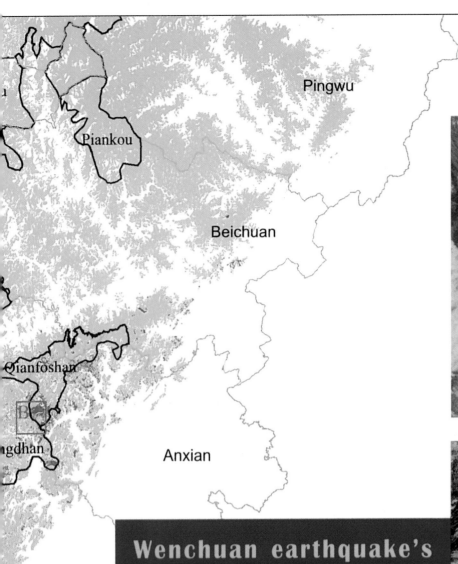

0 250 500 1000m

Collapsing high-voltage wire
transmission towers

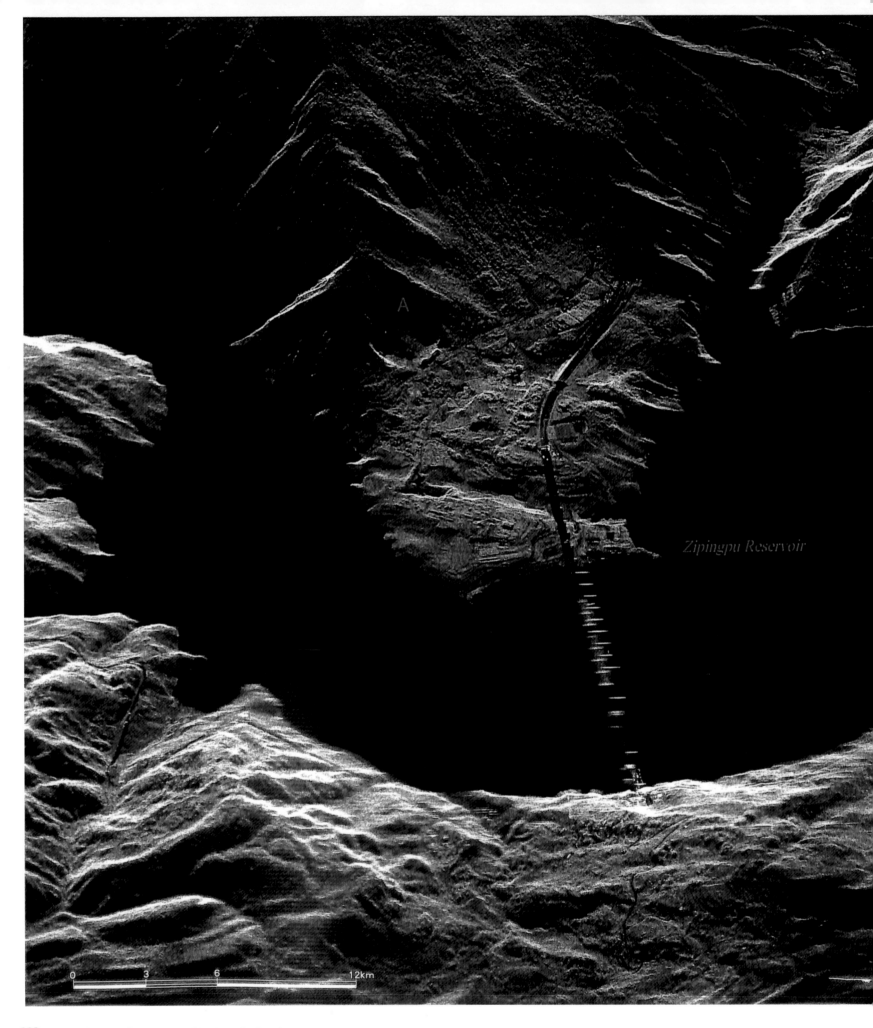

A

Zipingpu Reservoir

0 3 6 12km

Airborne remote sensing image of Erwang Temple acquired after the earthquake (on May 18, 2008)

Picture of Erwang Temple after the earthquake

Erwang temple built in memory of Li Bing and his son for their great contribution in leading the construction of the Dujiangyan Irrigation System

Two main halls built during the Qing Dynasty remained mostly intact, however, the front gate, the side halls, the slope protection works, and the rocky mountain paths, all built in 1990s, were badly damaged.

IKONOS satellite image acquired before the earthquake (on September 19, 2007)

Airborne remote sensing image acquired after the earthquake (on May 18, 2008)

The Baopingkou Diversion Passage and the Feishayan Floodgate were unaffected

Juyuan High

0 50 100 200m

▲ Airborne optical image of the town of Juyuan, in the city of Dujiangyan

Acquired May 18, 2008. The town of Juyuan is no more than 25 km away from the town of Yingxiu. The houses in the town of Juyuan were also seriously damaged, and over one-fifth of them collapsed. The red frame inside the image marks Juyuan High School, where heavy casualties occurred.

◀ SAR image on May 15, 2008 also shows collapsed houses and buildings in the city of Dujiangyan. Examples include Xinjian Elementary School and the Traditional Chinese Medicine Hospital of the city of Dujiangyan.

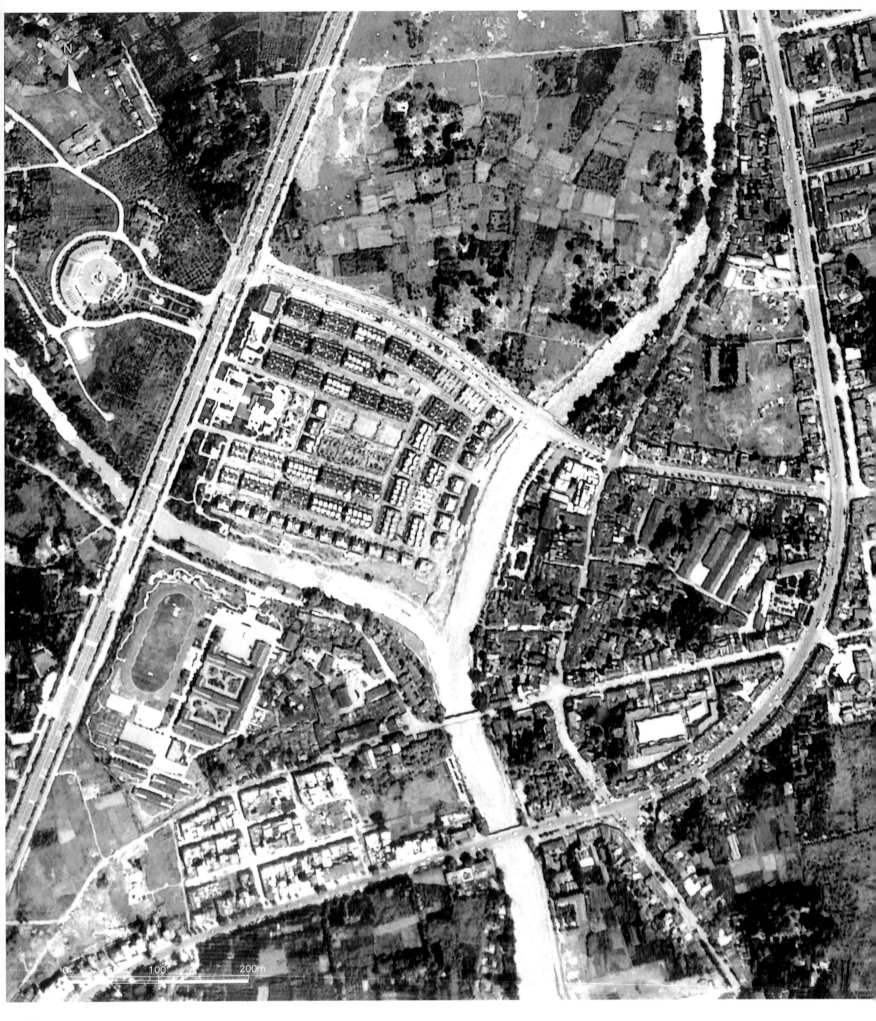

◄ **Airborne optical image of the town of Zhongxing, in the city of Dujiangyan**

Acquired May 19, 2008. The town of Zhongxing is 25 km away from the town of Yingxiu. The houses there were heavily damaged, with about 20% of houses in the town collapsed, and conditions in the countryside were much worse.

Distribution of the collapsed buildings in Maoxian County

Maoxian County is located in the southeast section of Aba Tibet and Qiang Autonomous Prefecture, which is the transitional zone from the Qinghai-Tibet Plateau to the Sichuan Plain, and is the main residential area of the Qiang people. Maoxian County lies along the Wenchuan-Maoxian seismic belt, and nearly 20% of the houses were either severely damaged or collapsed in the urban area of the Maoxian County.

Legend

Collapsed houses

Road

River

Acquired May 23, 2008. Maoxian County lies on the Wenchuan-Maoxian Fault. In the urban area of the county, about 20% of the buildings were either collapsed or severely damaged, and these collapsed buildings can be seen clearly on the *IKONOS* image. The red frame in the image represents the most severe damage.

▼ Airborne SAR image of Maoxian County

Comparing post-quake X-band high resolution SAR image (acquired May 14, 2008) with pre-quake 2.5-m resolution *SPOT* image (acquired November 11, 2005), some collapsed houses in both rural and urban areas were found.

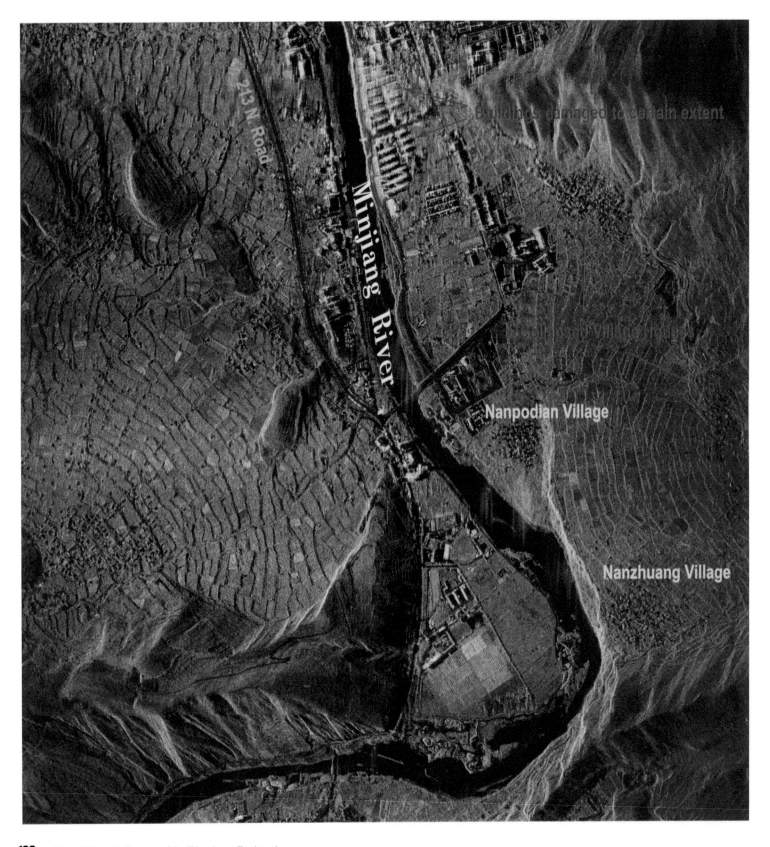

▼ Pre-earthquake *SPOT 5* image of Maoxian County (acquired November 1, 2005)

Distribution of the collapsed houses in the town of Zagunao, in Lixian County

Lixian County, situated in western Sichuan Province, has a population of more than 40,000 people, most of whom are Tibetan or Qiang. The town of Zagunao, which means "lucky place" in Tibetan, is in the urban district of Lixian County, located in mountain valleys about 54 km away from the epicenter town, Yingxiu. More than 20% of houses in Zagunao were either collapsed or seriously damaged by the earthquake.

Legend

Collapsed houses

Road

River

0 50 100   200m

Airborne optical images of the town of Shigu, Maoxian County

Acquired May 15, 2008. The town of Shigu is located on the east side of the Minjiang River. The houses and the buildings of the town of Shigu were relatively slightly damaged by the earthquake, and fewer than 20% of them collapsed. The red frame within the image shows the location of the Jiyu power station of the Baoshan Group, which was at high risk due to the earthquake and the earthquake-induced landslides.

Acquired May 19, 2008. The town of Longfeng, located in the piedmont plain of Longmen Mountain, is famous for producing garlic. It is 6 km away from the Anxian-Guanxian Fault and 19 km away from the Yingxiu-Beichuan Fault. About 20% of the houses in the town of Longfeng collapsed.

Acquired May 16, 2008. The town of Dabao is located in the mountainous area of the city of Pengzhou. Since the Yingxiu-Beichuan Fault passes through the town of Dabao, its collapse rate was higher than 90%. No effective disaster-relief could access this area because of a traffic block caused by earthquake-induced landslides. The red frame within the image shows the collapsed houses.

Acquired May 19, 2008. The town of Hongbai lies right above the Yingxiu-Beichuan Fault, so its collapse rate was higher than 90%. The red frame shows collapsed houses and disaster-relief tents.

▼ Airborne optical image of the village of Xiaojiaping, in the town of Longmenshan, in the city of Pengzhou

Acquired May 23, 2008. The village of Xiaojiaping is located in the town of Longmenshan, in the city of Pengzhou. The Yingxiu-Beichuan Fault passes through the town of Longmenshan. Therefore, the house collapse rate was higher than 90%.

▲ Airborne optical image of the town of Yinghua, in the city of Shifang

Acquired May 19, 2008. The town of Yinghua is located in the area between the Yingxiu-Beichuan Fault and the Anxian-Guanxian Fault. The collapse rate was about 75%, and the collapse in the north part of the town was more serious than that in the south part.

Acquired May 23, 2008. The town of Yunxi is 4 km away from the Anxian-Guanxian Fault, and had a collapse rate of about 20%. The red frame in the image shows collapsed and damaged houses.

Distribution of collapsed houses in the city of Mianzhu

The city of Mianzhu is located in the northwest part of the Sichuan Basin, and has an area of 1,245 km^2, and ranges in altitude from 504 m to 4,406 m. This city is typically characterized as "Six parts hill land, three parts farm land, and one part water." The Yingxiu-Beichuan Fault and the Anxian-Guanxian Fault pass through this city, so damage was serious and collapse rates were around 40%. The collapse rate increased from the southeast to the northwest, and the collapsed houses were mainly distributed in the nine towns between the town of Guangji and the town of Gongxing. The damage in rural regions was more serious than that in urban areas. The collapse rate in urban regions was less than 10% on the earthquake fault from the town of Guangji to the town of Gongxing. The collapse rate increased to 25% in villages and towns, while exceeding 50% in rural regions.

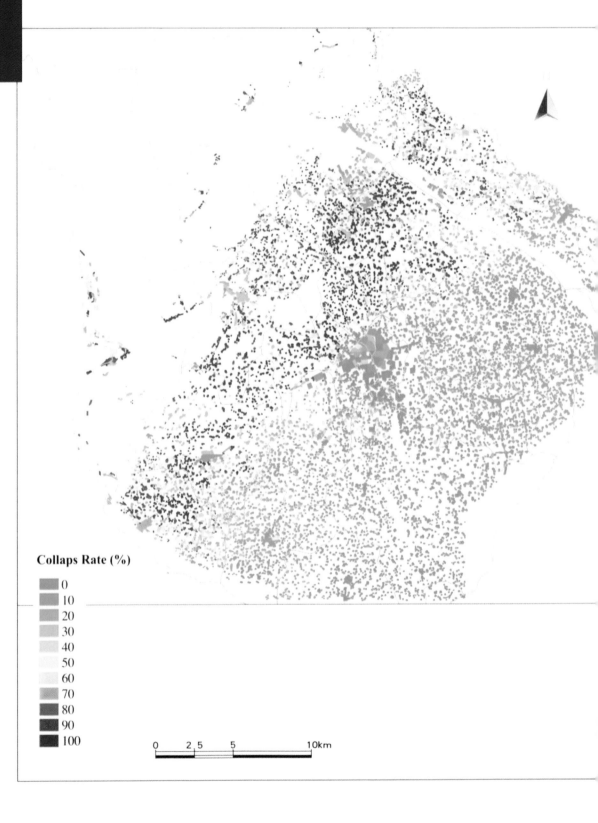

Collaps Rate (%)

0
10
20
30
40
50
60
70
80
90
100

0   2.5   5   10km

Hanwang Town

▲ Radar image in the town of Hanwang, in the city of Mianzhu

Acquired May 24, 2008. The radar image shows that the collapse and destruction was serious and some adjacent houses collapsed together.

▲ Houses collapsed or destroyed by the earthquake in the town of Hanwang

◄ Airborne optical image of the town of Hanwang, in the city of Mianzhu

Acquired May 19, 2008. The town of Hanwang is between the Yingxiu-Beichuan Fault and the Anxian-Guanxian fault. The damage was extremely serious, about 50% of the houses were collapsed, and more than 20,000 people were killed in this disaster. (A) Collapsed and destroyed houses (B) Destroyed Dongfang Turbine Factory.

Acquired May 19, 2008. The town of Guangji is in the piedmont transition region of Longmen Mountain and between the Yingxiu-Beichuan Fault and the Anxian-Guanxian Fault. Approximately 75% of the houses in the town of Guangji collapsed. The red frame shows collapsed and seriously damaged houses.

0 100 200   400m

▲ Collapsed building in the town of Guangji

► Airborne optical image of the town of Zundao, in the city of Mianzhu

Acquired May 19, 2008. The town of Zundao is in the piedmont transition region of Longmen Mountain, and between the Yingxiu-Beichuan Fault and the Anxian-Guanxian Fault. The houses in the town of Zundao were seriously damaged, and the collapse rate approached 75%. The red frame shows the collapsed and damaged houses.

Acquired May 19, 2008. The town of Guangji is in the piedmont transition region of Longmen Mountain and between the Yingxiu-Beichuan Fault and the Anxian-Guanxian Fault. Approximately 75% of the houses in the town of Guangji collapsed. The red frame shows collapsed and seriously damaged houses.

▲ Collapsed building in the town of Guangji

0   100   200   400m

▶ Airborne optical image of the town of Zundao, in the city of Mianzhu

Acquired May 19, 2008. The town of Zundao is in the piedmont transition region of Longmen Mountain, and between the Yingxiu-Beichuan Fault and the Anxian-Guanxian Fault. The houses in the town of Zundao were seriously damaged, and the collapse rate approached 75%. The red frame shows the collapsed and damaged houses.

Distribution of the destroyed houses in the town of Anchang, in Anxian County

The town of Anchang is an old urban district of Anxian County; Longmen Mountain is northwest of the town of Anchang. Topography southeast part of the town of Anchang features plains and hills, and landslides there were not severe. In the town of Anchang, in the Anxian-Guanxian Fault, almost 20% of the houses collapsed, while in the country the situation was much worse.

Legend

- Collapsed houses
- Road
- River

0   125   250   500m

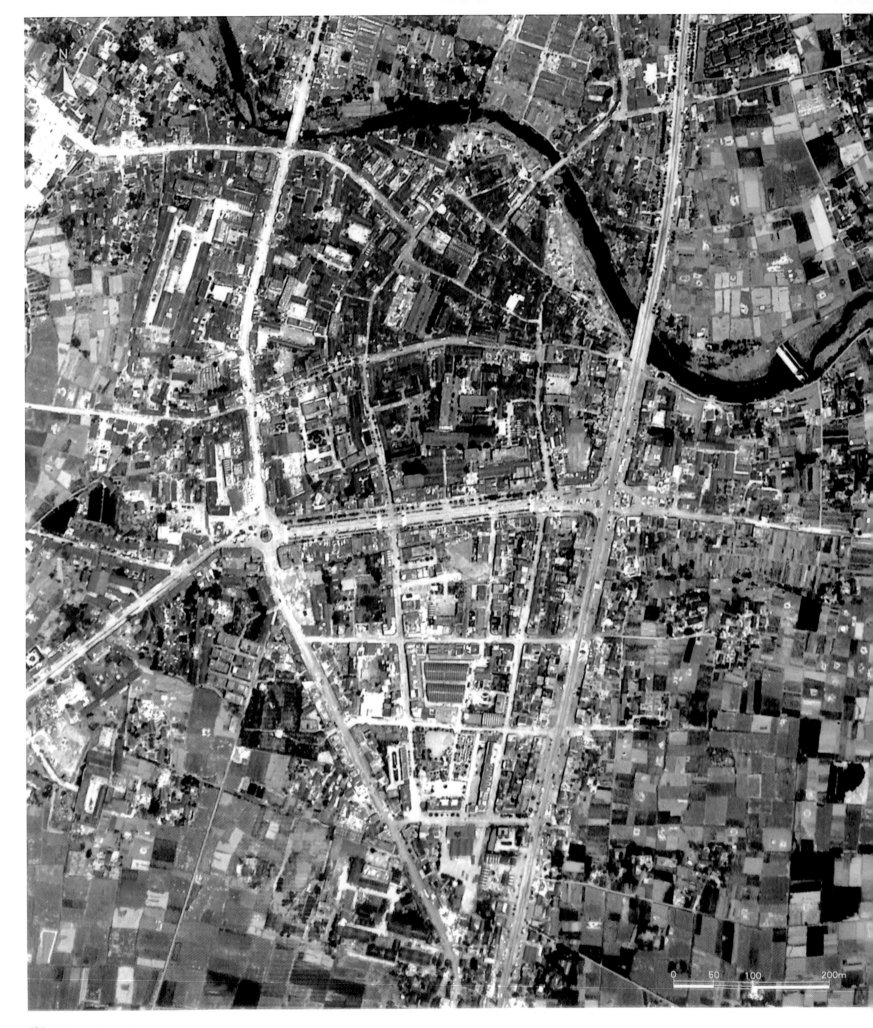

Acquired May 19, 2008. The town of Xiushui, located in the Anxian-Guanxian Fault, is the largest town in northwest Sichuan, with a population of more than 60,000. Houses were seriously damaged, at a rate of more than 60%.

▼ Airborne optical image of the town of Xiaoba, in Anxian County

Acquired May 19, 2008. Xiaoba is located between the Yinxiu-Beichuan Fault and the Anxian-Guanxian Fault, so houses were heavily damaged; the collapse ratio was more than 60%.

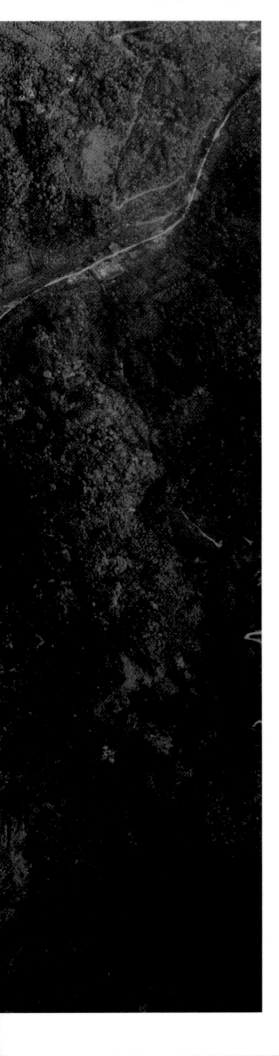

◄ **Airborne optical image of the town of Chaping, in Anxian County**

Acquired May 23, 2008. The Yingxiu-Beichuan Fault just cuts across the town of Chaping, so houses were seriously damaged, with a building-collapse ratio of more than 80%. The red frame delineates destroyed houses.

Distribution of destroyed houses and buildings in the city of Mianyang

Maoxian County is located in the southeast section of Aba Tibet and Qiang Autonomous Prefecture, which is the transitional zone from the Qinghai-Tibet Plateau to the Sichuan Plain, and is the main residential area of the Qiang people. Maoxian County lies along the Wenchuan-Maoxian seismic belt, and nearly 20% of the houses were either severely damaged or collapsed in the urban area of the Maoxian County.

Legend

⊡⊡ Collapsed houses

Road

River

0 0.25 0.5 1km

Acquired May 27, 2008. In this image, destroyed houses can be seen clearly.

Post-earthquake and pre-earthquake *Radarsat* images of the city of Mianyang

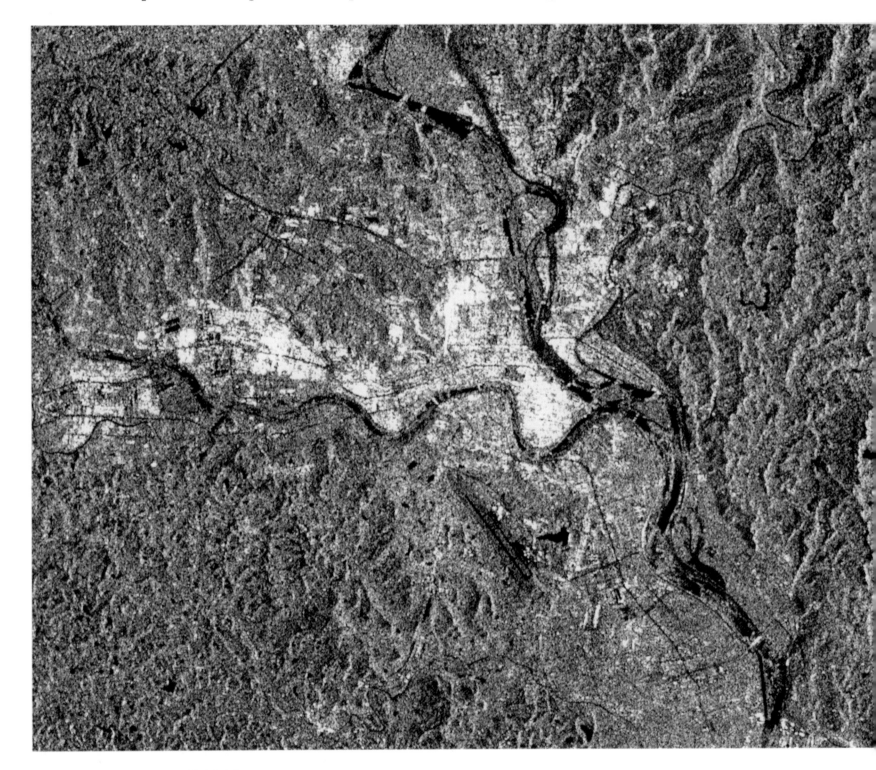

Post-earthquake image: May 14, 2008

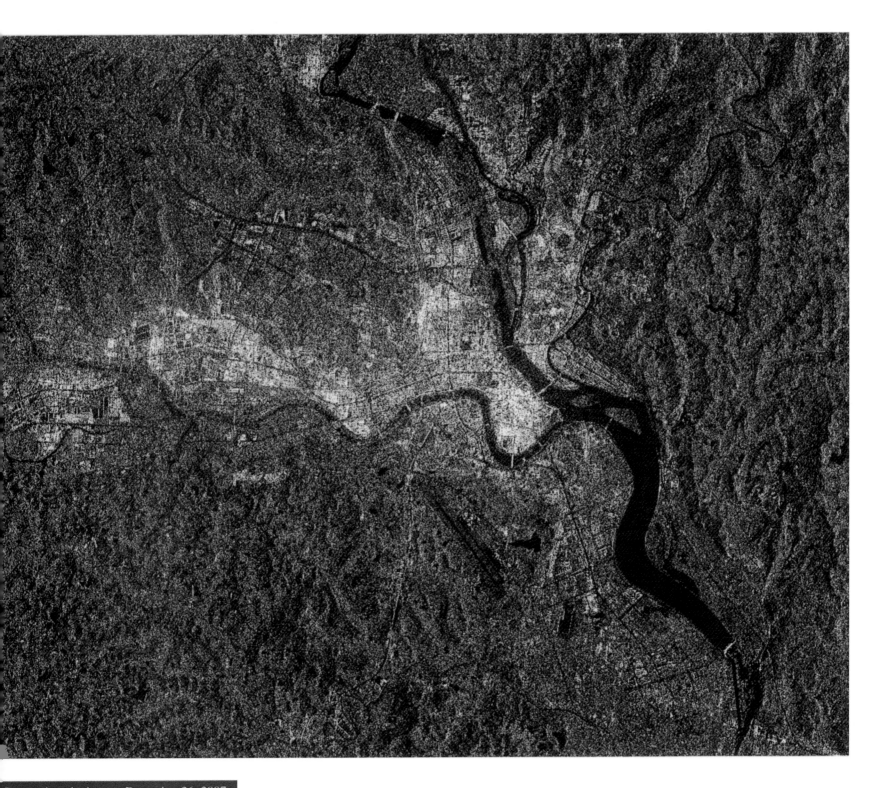

Pre-earthquake image: December 26, 2007

▲ Airborne optical image of the town of Xinzao, in the city of Mianyang

Acquired May 18, 2008. The town of Xinzao, a rural town in the city of Mianyang, was quite seriously affected. The house collapse rate was more than 40%, and was higher than the collapse rate for the factories and the other buildings. The red frame within the image shows the collapsed and severely damaged houses and buildings.

▶ Airborne optical image in the town of Qingyi, in the city of Mianyang

Acquired May 18, 2008. The town of Qingyi is another heavily hit rural town in the city of Mianyang. The rate of collapsed houses was more than 20%, and was higher than the rate of collapsed factories and other buildings. (A) The new campus of the Southwest University of Science and Technology. There are slight property losses, that is, some buildings were damaged, but did not collapse.

◄ **Airborne optical image of the town of Caopo, in Wenchuan County**

The town of Caopo, located 20 km away from Wenchuan County, was affected by large landslides, and therefore highways were destroyed. The town had become an isolated island, and the message "SOS700" was found on a roof.

◄ *WorldView* image of the town of Wujia, in the city of Mianyang

In this image, acquired May 16, 2008, which compares a post-earthquake *WorldView* satellite image with a high-resolution airborne optical image, some collapsed houses can be identified by the *WorldView* panchromatic band. But it is harder to identify collapsed houses in this image than it is with an airborne image under the same resolution.

Distribution of collapsed houses in the city of Jiangyou

The city of Jiangyou is located in the northwest part of Sichuan Basin, along the upper reaches of the Fujiang River, southeast of Longmen Mountain. It is the national metallurgy industrial base, and it is a vital energy resource base of Sichuan. The house collapse here was severe, with about 5% of buildings collapsed in the urban area of the city of Jiangyou.

Legend

◾ ◾ ◾ Collapsed houses

Road

River

0   200   400   800m

Post-earthquake and pre-earthquake *Radarsat* images of urban area of the city of Jiangyou

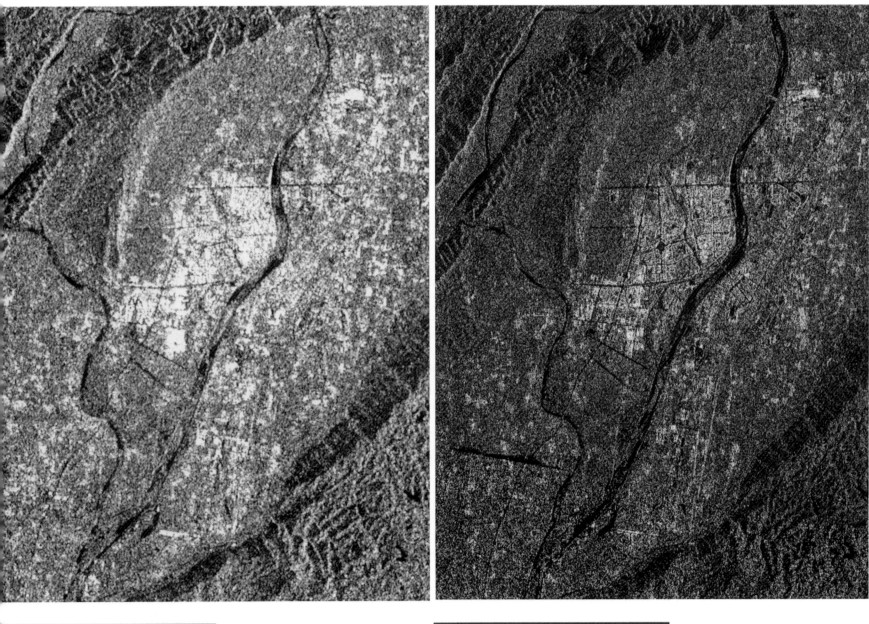

Post-earthquake image: May 14, 2008

Pre-earthquake image: December 19, 2007

▲   Airborne optical image of the town of Xiping, in the city of Jiangyou

Acquired May 18, 2008. The distance from Xinping to the urban area of Beichuan County is just 20 km. The houses in the town of Xinping were severely damaged, and the house collapse rate was over 25%.

▶   Airborne optical image of the town of Sandui, in the city of Guangyuan

Acquired May 18, 2008. The town of Sandui lies on the Yingxiu-Beichuan Fault but only a few houses, less than 10%, collapsed.

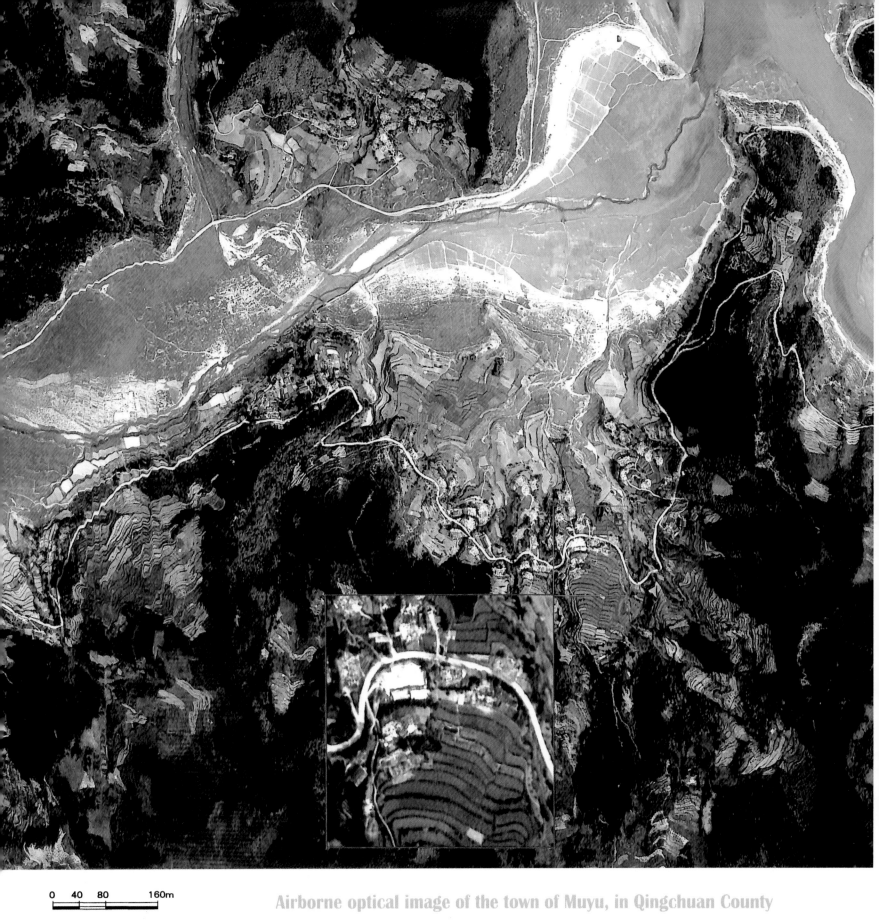

0 40 80 160m

Acquired May 18, 2008. The town of Muyu is in the northeast area of the Wenchuan-Maoxian Fault. The seismic intensity in Muyu-Shazhou was high, and the house collapse rate in this area was above 60%, which is higher than the surrounding area. (A) Collapsed houses in urban areas. (B) A village that was being razed.

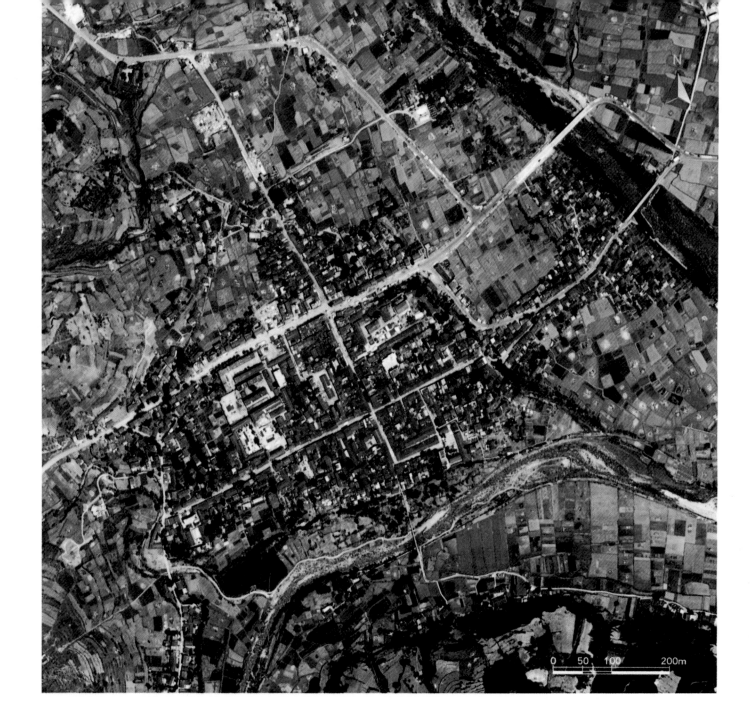

▼ Airborne optical image of the town of Qingxi, in Qingchuan County

Acquired May 28, 2008. The town of Qingxi, built more than 1700 years ago, has many historic sites. Although it is located on the Wenchuan-Maoxian Fault, building collapse here was less severe, with a collapse rate of about one-sixth.

◀ Airborne optical image of the town of Guanzhuang, Qingchuan County

Acquired May 28, 2008. The town of Guanzhuang is located between the Yingxiu-Beichuan Fault and the Wenchuan-Maoxian Fault. The houses in the town of Guanzhuang were severely damaged; more than 75% of them collapsed. (A) Collapsed or damaged houses. (B) Disaster-relief tents.

0   100   200   400m

▲ Airborne optical image of the town of Long'an, in Pingwu County

Acquired May 23, 2008. The urban area of Pingwu County is located in the town of Long'an. In the 1976 Songpan-Pingwu earthquake, losses were very heavy. Since much attention was paid to house construction when the area was rebuilt after the 1976 Songpan-Pingwu earthquake, and because the earthquake intensity in the town during the Wenchuan earthquake was low, losses were light, and very few houses collapsed.

Acquired May 28, 2008. The town of Nanba is located between the Yingxiu-Beichuan Fault and the Wenchuan-Maoxian Fault. More than 80% of the houses in the town of Nanba were severely damaged or collapsed. The red frame in the figure shows Nanba Primary School. Two three-story buildings of this school completely collapsed, and many teachers and students in this school died. The heroic actions that 48-year-old teacher Du Zhengxiang took in saving her students moved the whole country.

▲ Airborne optical image of the town of Chenjiaba, in Beichuan County

Acquired May 28, 2008. In Chenjiaba, a town crossed by the Yingxiu-Beichuan Fault, the houses were heavily damaged, with over 80% of them destroyed. (A) A landslide split the town into two parts and caused heavy casualties. (B) Collapsed houses.

▶ *IKONOS* image of the town of Yuli, Beichuan County

Acquired May 23, 2008. The image shows that the floodwater of the Tangjiashan Barrier Lake began to submerge the town of Yuli.

Post-earthquake and pre-earthquake *Radarsat* images of urban area of the city of Deyang

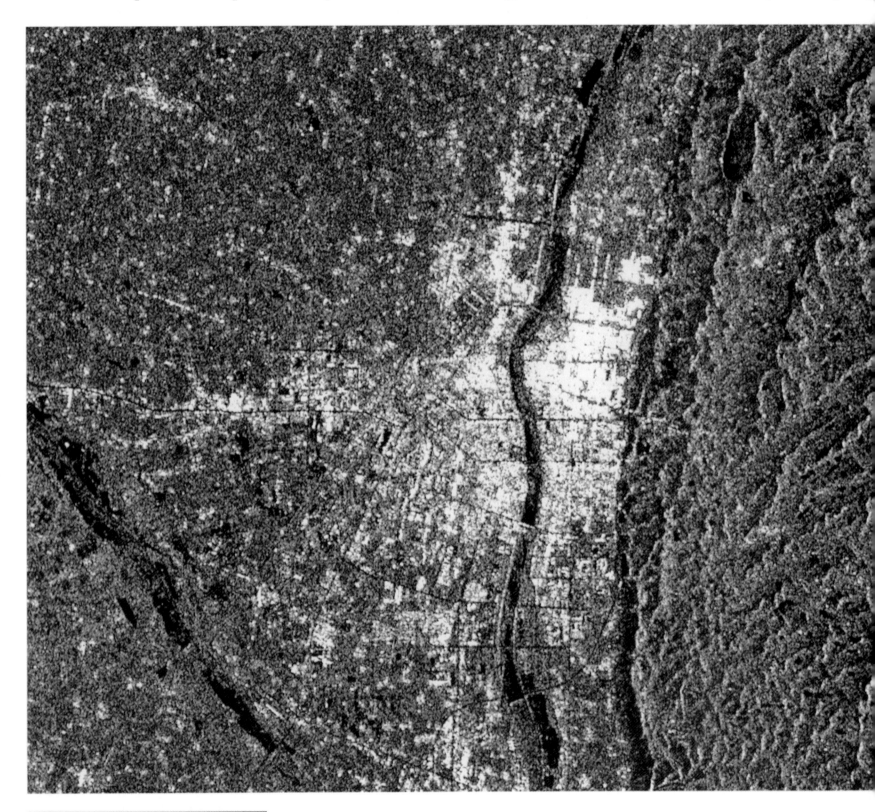

Pre-earthquake image: December 18, 2007

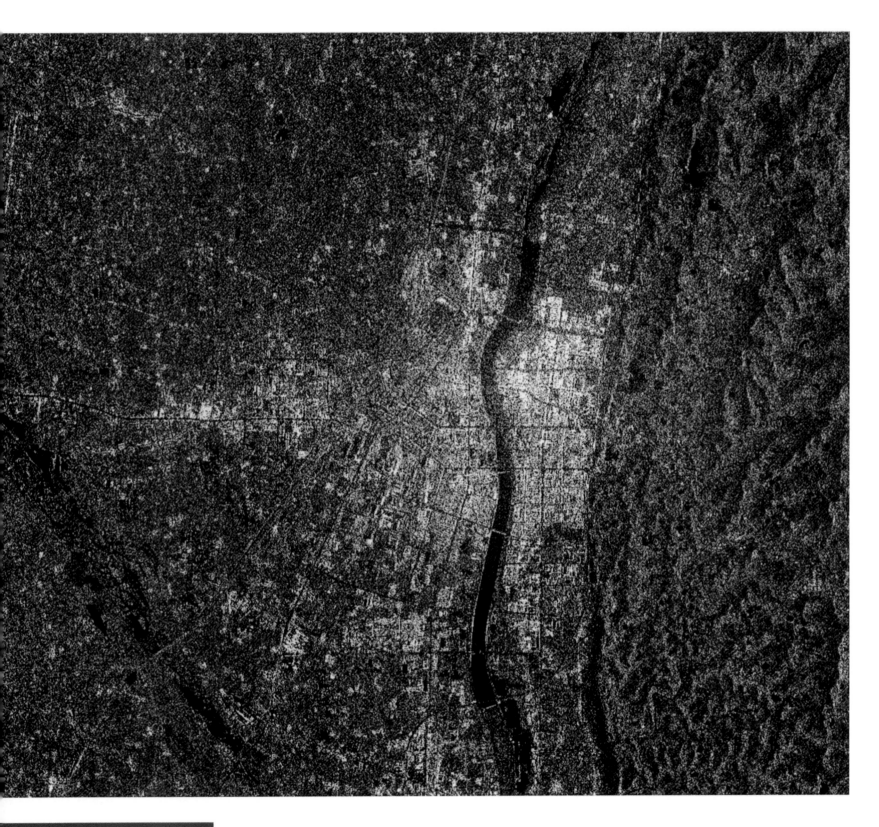

Post-earthquake image: May 14, 2008

Acquired May 16, 2008. The town of Yuli is the hometown of Da Yu and is located between the Yingxiu-Beichuan Fault and the Wenchuan-Maoxian Fault. Houses and buildings in this town were severely damaged, over 60% of the buildings collapsed. "SOS" did not show up in the farmland of the town of Yuli in the red frame of this image until 19 May.

0   50   100   200m

▲ Airborne optical images of the town of Yuli, in Beichuan County

Acquired May 27, 2008. The floodwater of Tangjiashan Barrier Lake inundated some houses in the town of Yuli. (A) The flooded area. (B) The "SOS" sign marked by local residents.

▲ Airborne optical image of the town of Leigu, in Beichuan County

Acquired May 19, 2008. The town of Leigu is located on the Yingxiu-Beichuan Fault. Over 75% of the houses were destroyed due to the earthquake. Since the town of Leigu has relatively wide open spaces, it became a temporary settlement for disaster affected people from Beichuan County and a transfer base for relief materials. Many relief tents are visible in this image (see the red frame).

▲ Airborne optical image of the town of Leigu, in Beichuan County

Acquired May 23, 2008. As a post-quake relief settlement and transfer base for relief materials in Beichuan County, more relief tents appeared on this image, which was taken on May 23, 2008 (see the red frame).

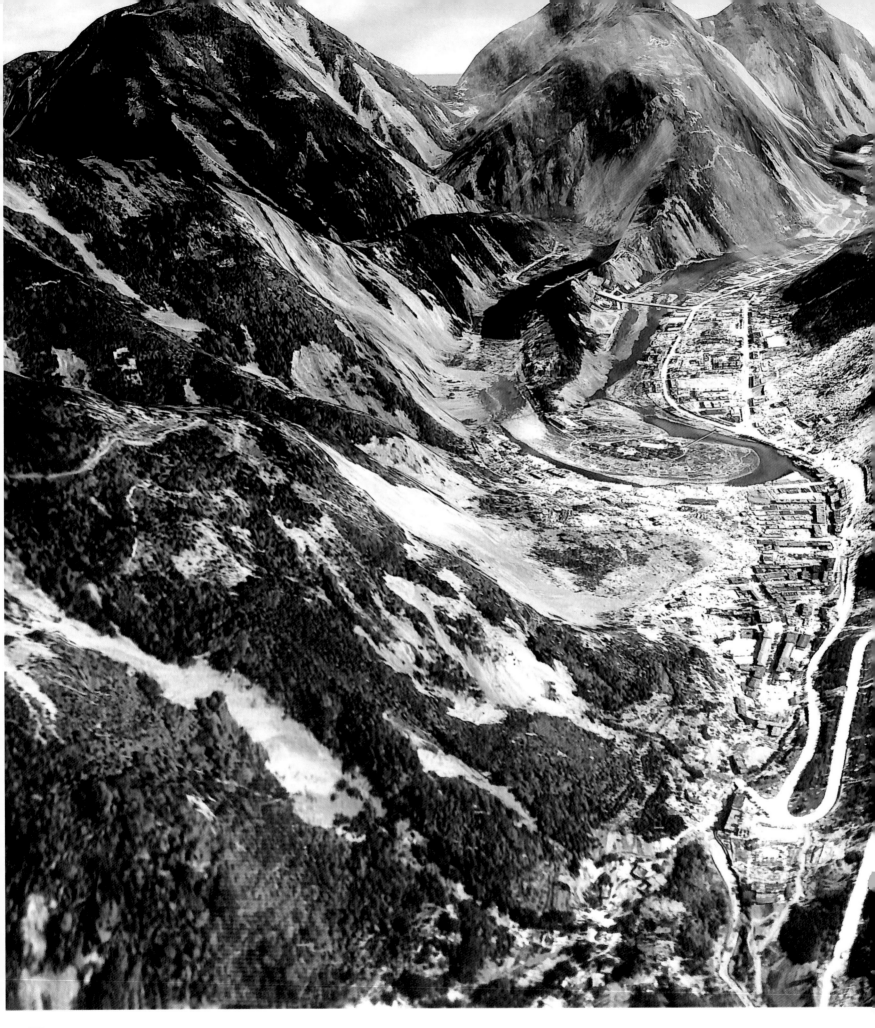

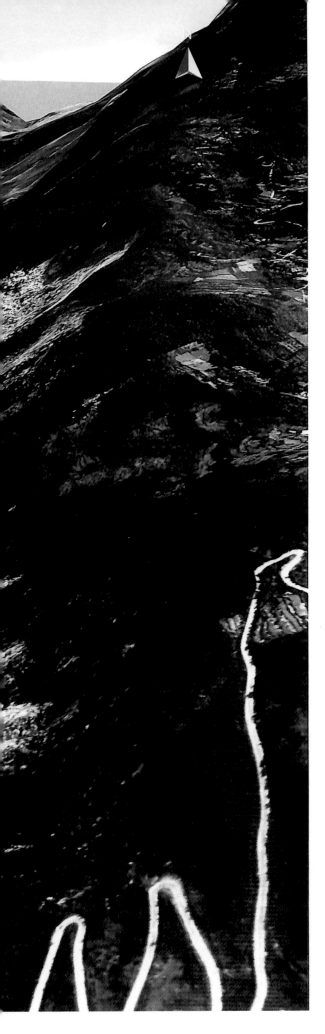

▲ The above photo shows the scene of most parts of the town of Qushan before the earthquake. The lower image is the old urban area of the Beichuan County, which suffered severely during the earthquake.

Three-dimensional airborne remote sensing image of Beichuan County

◄ From the three-dimensional remote sensing image, we can see that the May 12 earthquake not only destroyed buildings, but also changed the topography and configuration of surrounding mountains. Wide-ranging landslides occurred in these mountains. Many houses collapsed not because of the earthquake, but rather because of landslides. Those landslides, and the falling rocks that accompanied them, caused buildings to collapse and caused many people to lose their lives.

The airborne optical image of Beichuan County

◄ Beichuan County is located between two mountains: one is an offshoot of the Himalayas, and another is an offshoot of Longmen Mountains. One side is a lively, multi-alley, old urban district; another side is a quiet, comfortable, new urban district. All the buildings in the urban district were constructed against the hill and along the river. Longmenshan Fault runs northwest beginning in Qingchuan County in the north, then westward past Beichuan County, Maoxian County, Dayi County, and near Luding County. The earthquake fault just passes through the urban area of Beichuan County; therefore, the May 12 earthquake caused catastrophic destruction to Beichuan County. More than 80% of the houses in the urban area of Beichuan County collapsed, and heavy casualties occurred. Meanwhile, secondary geological disasters such as barrier lakes and landslides also occurred.

Airborne remote sensing classification image for collapsed houses in the old urban district of Beichuan County

We used airborne ADS40 optical image (acquired May 16, 2008) of collapsed houses in the old urban district of Beichuan County to classify buildings such that the pink regions represent collapsed houses and red regions represent houses that did not collapsed during the earthquake.

◄ Airborne optical image Beichuan County, downtown (acquired May 16, 2008)

Three-dimensional airborne image of an urban area of Beichuan County (partial) (acquired May 27, 2008)

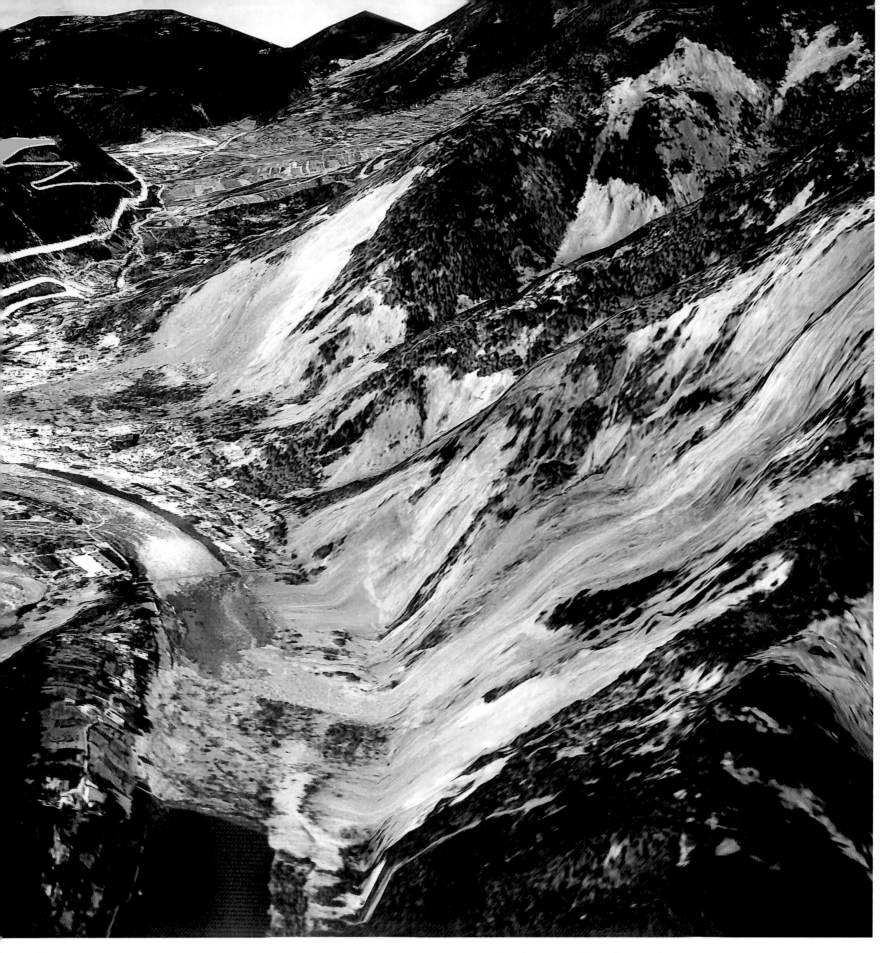

An urban district of Beichuan County was almost totally razed after the earthquake. Huge landslides buried buildings and caused extremely heavy casualties.

Three-dimensional airborne image of the old urban district of Beichuan County (acquired May 27, 2008)

The old urban district in Beichuan after the earthquake

The old urban district is located in the foothills of the Wangjiayan Mountains. The large-scale landslides induced by the earthquake caused complete burial of several streets near the foothills.

Three-dimensional airborne image of the new urban district of Beichuan County (acquired May 27, 2008)

New campus of Beichuan Middle School

Administration building of Beichuan County

Front of Beichuan Grand Hotel

The new urban district in Beichuan County is located near Jingjia Mountain. The landslides mercilessly buried Beichuan Middle School and many surrounding buildings, causing heavy casualties.

Chapter 5

DAMAGED ROADS

This chapter presents images of badly damaged roads and estimates of the amount of road damage. Images used are from Airborne ADS 40 and Airborne SAR and elsewhere. The images were selected to show the massive damage caused by the disastrous earthquake, as well as the secondary disasters that followed.

Using mainly airborne ADS40 data, we estimated road damage as follows: First we semi-automatically digitized the four road types, national roads (NR), provincial roads (PR), county roads (CR), and village roads (VR). Then we marked road damage by section with five different damage grades. The damage grades are hardly damaged, blocked by rolling rocks, covered by earth and rocks, roadbed ruined, and flooded. There are also four attributions for bridges, hardly damaged, ruptured, fallen, and inundated. On qualitatively attributed roads, statistical analysis was done, and thematic charts were mapped. The ADS40 images do not cover all the heavily damaged areas, and not all the areas covered are heavily damaged. Since the main roads and the most damaged roads were almost all along rivers and on hillsides, the estimation was not carried out on the whole affected area, but only in those areas that were seriously damaged.

Our damage analysis concluded that the length and grade of damaged roads are highly correlated to the northeast-trending Longmenshan Fault, so the roads near the fault between Yingxiu and Beichuan as well as the roads between Xuankou, Dujiangyan, and Maoxian were highly damaged, with longer sections and higher grades more likely to be damaged. Also, roads along the steep Minjiang River and Jianjiang River valleys were more heavily damaged. Some flooded sections located along the river valley were inundated by the dammed lakes. Statistical analysis was also done with respect to county and road hierarchy. The result shows that although there are fewer roads in the mountainous area than in the plains area, road damage was heavier in the mountains. In the area covered by the ADS40 images, with Longmen Mountain as a boundary, the west and northeast areas were more damaged by the earthquake and its after effects, mostly geological disasters, than the east and southeast part, where the roads were also less damaged. The damaged national roads are the sections of NR317 around Lixian County and NR213 around Wenchuan and Maoxian counties. Roads along the Laisuhe River banks, as well as the Minjiang and Caobahe rivers, were seriously affected by landslides and debris flow; up to 40% of these roads are damaged. When broken down by county, Wenchuan saw the highest damage. Amazingly, NR108, which passes though the city of Mianyang, was hardly affected. Provincial road 303 passing through Wenchuan County; PR302 passing through Beichuan County, Maoxian County, and the city of Jiangyou; and PR105 in Qingchuan and Pingwu counties were all badly damaged, which again follows the trend along the fault. This trend is also seen with county roads and village roads, which are more heavily damaged near the fault than those far away from the fault.

Out of 300 bridges, 49 were ruined. Destroyed bridges were located in Lixian County, Wenchuan County, and along both sides of the Longmenshan Fault, where more of the damaged bridges collapsed completely than cracked or broke in parts. The results of our statistical analysis show total damage length by different grades: 125 km of roads were piled with rocks; 233 km of roads were covered by landslide debris; 128 km of roadbeds were ruined; and 19 km of roads were flooded.

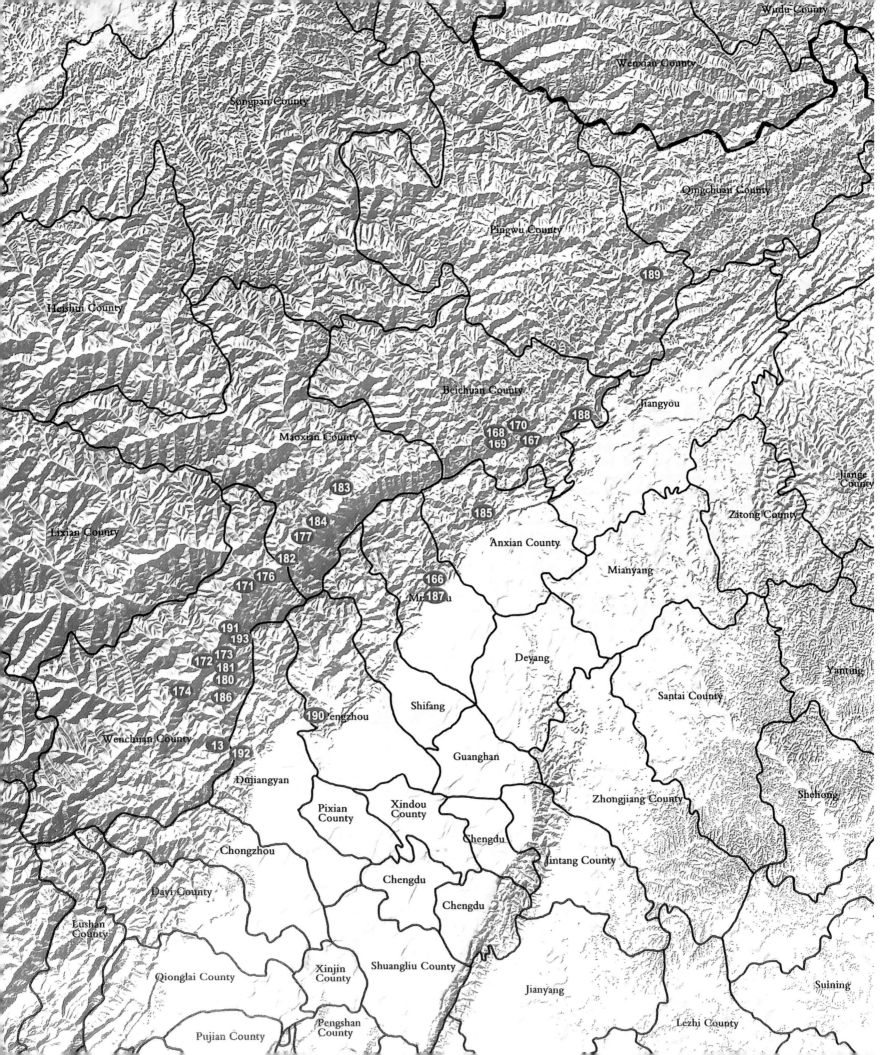

▲ Airborne ADS40 image of the village of Xiejunmen, in the town of Tianchi, in the city of Mianyang

Airborne ADS40 image acquired on May 16, 2008 shows that the earthquake caused quite a few landfalls and landslides, which then blocked nearly 30% of the roads in this image. From (A) one can see huge rocks blocking the road at the bottom of the image. On the west bank of the Mianyuan River (B) and (C) dammed lake submerging more than 300 m of roads along the Mianyuan River. (D) Heavy landslides on one side of the river even affected the road on the other side.

P. Road 302

Jianjiang River

0   50   100   200m

▲ Airborne ADS40 image of Jianjianghe highway, in Beichuan County

This airborne ADS40 image acquired on May 19, 2008 shows a section of Provincial Road 302 in Beichuan County was seriously damaged, particularly in the Jianjiang Road area. The 667 m Jianjiang River road was composed of (A) Xiayu Bridge, of which about 50 m in the middle of which partially dropped down. (B) Longweishan Tunnel and (C) Shisuoyi Bridge. Over 65% of section A collapsed, and the road was fractured to the west, marked (D). The near side of the entrance of Longweishan Tunnel was blocked by landfall. A section of Provincial Road 302 (E) fell down and (F) distorted.

◀ Picture of Xiayu Bridge taken after the earthquake

▲ Airborne ADS40 image of the village of Zhicheng, in the town of Yuli, in Beichuan County, May 16, 2008

On the airborne ADS40 image acquired on May 16, 2008 (the left one), the roads were visible, although there was a dammed lake in the lower Jianjiang River. But the water level kept rising and the town of Yuli was flooded afterward. (This area was in danger of flooding when the image was acquired 19 May 2008, which is not shown here). The image on the next page was acquired on May 27, 2008, when the government began to discharge from the dam holding the lake.

▲ Airborne ADS40 image of the village of Zhicheng, in the town of Yuli, in Beichuan County, May 27, 2008

This image was acquired on May 27, 2008, when the government began to discharge from the dam holding the lake. However, the roads of the village of Zhicheng were still under water, including (A) a bridge, (B) a 1700-m-long section of Provincial Road 302, as well as (C) and (D) some other roads, where the flooded section (C) was 1700 m long, and (D) 1300 m. An "SOS" message was found on the image from May 18, 2008 (not shown here), which indicated that traffic in and out of the area was blocked, and as a result, food and supplies would need to be airlifted in.

▲ Airborne ADS40 image of the village of Zhangjiaba, in the town of Xuanping, in Beichuan County

This ADS40 image was acquired on May 16, 2008, at the same time as the image of the town of Yuli on the previous page. This area is a bit downstream of the town of Yuli, so it was flooded on May 16. Almost the whole section of Provincial Road 302 in this area was completely inundated, see (A). Other roads along the Jianjiang River were flooded as well, for example, (B), (C). Still some sections like (D) were obstructed by landfalls, though not really affected by flood.

► Airborne ADS40 image of the town of Guozhupu, in Wenchuan County

This ADS40 image was acquired on May 15, 2008. To the south of Wenchuan County, many landfalls, landslides, and debris flows took place, destroying quite a few roads. Some sections of National Road 213 were obstructed, such as (A) and (B) labeled on the image. (C) shows a more than 1400-m-long section of county road that was ruined by heavy debris flow.

P. Road 302

Minjiang River

A

B

C

0　100　200　400m

▲ Airborne ADS40 image of the town of Caopo, in Wenchuan County

From the ASD40 image acquired on May 15, 2008, numerous debris flows and landslides occurred along the Caoba River, as well as in the town of Caopo, which is a town with a population of about 4,000. These secondary geological disasters caused by the earthquakes badly damaged more than 80% of the roads along the river banks, marked (A), (B), (C), and (D) on the image. Some of the roads, which did not seem to be badly damaged, were obstructed by rocks or earth rolling from the adjacent mountain, such as the road around (E). The traffic of the town of Caopo did not recover until the end of June, so during this period, all daily supplies and groceries had to be sent by air or by foot.

▼ Airborne ADS40 image of the village of Zaojiaotuo, in the town of Yinxing, in Wenchuan County

This ADS40 image was acquired on May 16, 2008. It shows a bridge over the Minjiang River near the village of Zaojiaotuo that fell down as a direct result of the earthquake, marked (A). The east end of the bridge was broken, and there is a visible crack west of the bridge. (B) 280 m of heavily damaged county road, which was connected to the bridge. The road was almost completely covered by landfalls and debris flows, which are secondary geological disasters resulting from the earthquake. Because it is mountainous nearby and the slopes are steep (seen in this ADS40 image), geological disasters are overwhelming. West of the bridge, landfalls and landslides blocked more than 90% of National Road 317, marked (C). On the left corner of this image, the surface of the land is fragmented, hardly any vegetation or roads are visible.

◄ **Airborne ADS40 image of Taipingyi Reservoir, in Wenchuan County**

This ADS40 image acquired on 16 May 2008 shows severe damage of roads around Taipingyi Reservoir and the low-lying lands surrounding it. The upper part of the image shows the Taipingyi Reservoir, which was built from the Minjiang River. A bridge carrying National Road 213 across the Minjiang River, labeled (A), collapsed, and about 100 m fell down. East of the bridge, there were (B) heavy landslides and debris flows, which covered several sections of National Road 213. There were also landslides and debris flows on sections of National Road 213 (C) and (D) on the left bank of Minjiang River. Some of the other highways were also heavily affected: (E) the east bank of the Minjiang River is a case in point.

On this ADS40 image on May 15, 2008, National Road 213 travels along the right bank of the Minjiang River, which runs southwest around the area within this image. A huge landfall (A) occurred in the middle. The landslide made National Road 213 impassable for a length of 260 m. We do not know whether the roadbed beneath the landslide is passable, but considering the amount of weight placed on the roadbed by the slide material, it seems unlikely that the road underneath will be usable even after the debris is cleared. Along National Road 213, some medium landslides also brought debris and rocks that blocked the road. The road section marked (B) was covered by debris, which fell from the adjacent mountain. As we can see from the image, this debris has traveled rather far, moving along the valleys between mountains. (C) and (D) were about the same, except that road section (C) was a bit more heavily covered than (D). The bridge connected to (D) was broken, which is also visible from the image, and across the river, landslide or landfall is also visible.

◄ Airborne ADS40 image of Bingli Yanmen in Wenchuan

Acquired on May 15, 2008, this image shows damaged village roads. According to ancillary data from the local government, one of these roads had just been built to facilitate traffic between the small villages and between villages and towns, but the road foundations under several sections of the roads were ruined by landfalls and landslides, and some sections were covered with rocks and debris. The total length of damage was up to 1,360 m. As labeled, (A) was affected by a small landfall, which then blocked this section of the newly built road. The outer side the roadbed near (B) fell, and the road was blocked to traffic. Lands near (C) fell down and covered over 600 m of the new village road. (D) seemed less affected, except for some rocks blocking in the way and the falling of outer side, but it is doubtful that vehicles could even access this section of the road, since (A) is impassable. There is slight damage on section (E), compared to section (F), where the roadbed fell completely.

Post-earthquake photos of Baihua Bridge

Acquired on May 16, 2008, this ADS40 image shows a damaged section of National Road 213 along the west river bank in the village of Zhangjiaping, in the town of Yingxiu. As the image shows, about 100 m of the road was destroyed, marked (A). (B) shows a ruptured highway approach bridge ,with a visible crack. Landfalls and slides took place where the foundation was soft. (C), (D), and (E) show road sections that were blocked by rocks rolling from the landfalls or landslides nearby.

N

Minjiang River

E

A

B

N. Road 213

D

0   40   80   160m

▲ Airborne ADS40 image of the village of Fotangbagou, in the town of Yingxing, in Wenchuan County

Acquired on May 16, 2008, this ADS40 image shows the damage of National Road 317 around the village of Fotangbagou, in the town of Yinxing. (A) A 400-m-long section of National Road 317 was covered by debris from landslides. (B) and (C) Sections of county roads that were heavily covered by debris flow and landslides. It is estimated that over 70% of the road surfaces along the Minjiang River were blocked. As we see from the image, there are two or possibly more roads that were almost parallel along the east bank of the Minjiang River, of which the sections around (B) and (C) were all badly affected. Given the number and weight of the rocks, it is doubtful that the road underneath remains undamaged.

Acquired on May 16, 2008, this ADS40 image shows another section of National Road 317, a tunnel of which was blocked by landfall, marked (A). The image also shows that on the west bank of the Minjiang River the roads were totally covered as a result of unbelievably heavy landslides and debris flows. (B) and (C) Sections of the road were invisible, but since there is a bridge connected to it, there must have been roads or tunnel. The image also shows (D) huge rocks blocking the road that have a diameter of over 8 m.

Minjiang River

N. Road 213

0 40 80 160m

In the middle of this ADS40 image acquired on May 15, 2008, there was a chain of landslides along the east Minjiang River bank, which then caused destruction to National Road 213 for over 1,000 m. As shown on the image, (A), (B), and (C) marked the damaged sections, with the longest section (A) up to 230 m. The extent of the road blockage made quick repair impossible, as a result people were traversing the roadway by foot, but cars could not get through.

Airborne ADS40 image of the village of Yan'eryan, in the town of Fengyi, in Maoxian County

This ADS40 image acquired on May 15, 2008 shows the damage of National Road 213 along the west Minjiang River bank around the town of Fengyi. There were more landslides, landfalls, and debris flows on left side of the image than on the right. About 30% of the National Road 213 in this image was blocked or ruined, the most severley damaged areas are labeled (A), (B), and (C). The county road on the east river bank was affected as well; see sections marked (D) and (E).

As shown on this ADS40 image on May 15, 2008, on the right bank of Minjiang River a series of landslides and debris flows blocked National Road 213 for over 400 m. More than 60% of the small road around the village Jiyu was affected by debris flows. (A) and (B) Road sections of National Road 213 affected by geological disasters. (C) A section of country road covered by debris.

On May 19, 2008, more than 10 cars were seen to be stranded on the road around the village of Xiaojiaqiao, in the town of Qianfo, as a result of (A) disastrous landfalls. (B) Rocks densely littered over the road. (C) The landfalls also formed a dammed lake, which then flooded the road.

aojiaqiao Barrier Lake

C

A

0 40 80 160m

Along the Minjiang River banks, quite a few landfalls and debris flows occurred and brought great damage to National Road 317 and other county roads. As shown on this ADS40 image, which was acquired on 16 May 2008, up to 70% of National Road 317 was badly damaged, labeled (A), (B), (C), where the longest section (C) covered by debris was 1,440 m. The county road was also affected by landfalls; see (D). Even some sections not blocked had collapsed as a direct result of the earthquake or the fall of rocks nearby, as shown by (E), (F), and (G).

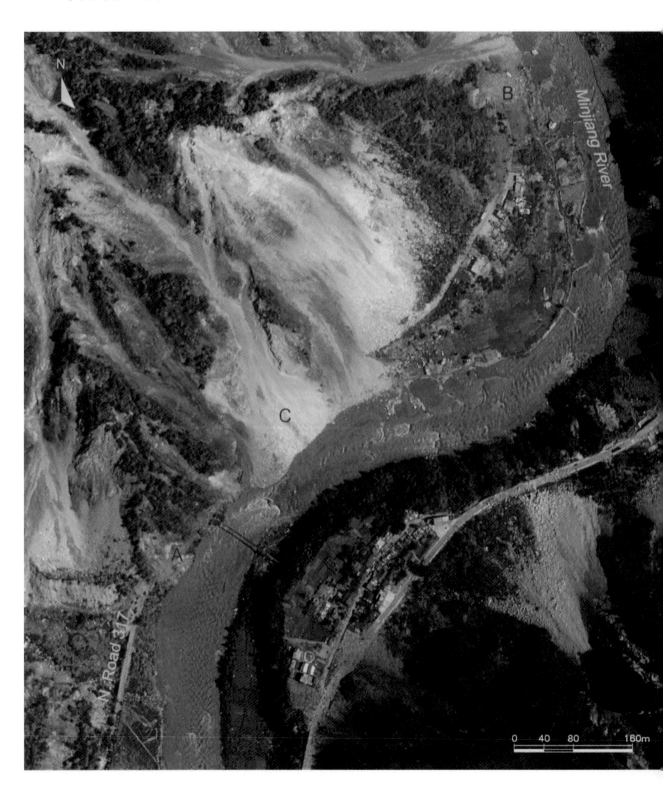

Minjiang River

▲ Airborne ADS40 image of the town of Hanwang, in the city of Mianzhu

This airborne image, acquired on May 19, 2008, shows that the railway in the northern part of the town of Hanwan was blocked by landslides in some sections. (A) and (B) Areas badly covered by landslides and landfalls. (C) One section was even affected by collapsed houses.

Distorted railway ▶

0 150 300 600m

▲ Airborne radar image over the village of Laochang, in the town
of Chenjiaba, in Beichuan County (May 25, 2008)

This image shows that there were several segments of Provincial Highway 302 destroyed by
landslides in the town of Chenjiaba, northwest of Beichuan County. (A) The area between the
villages of Xibahe and Laochang that was being threatened by the landslide along the road line. (B)
An area blocked by a large landslide, located between the villages of Laochang and Chenjiaba.

This image shows that provincial highways S105, S205, and their bridges in the town of Nanba were destroyed by the devastating earthquake and the debris flow. (A) A bridge of S205 that collapsed. (B) S105 blocked by landslide. (C) A large landslide where S105 was seriously damaged. (D) A collapsed bridge downtown (see enlarged image).

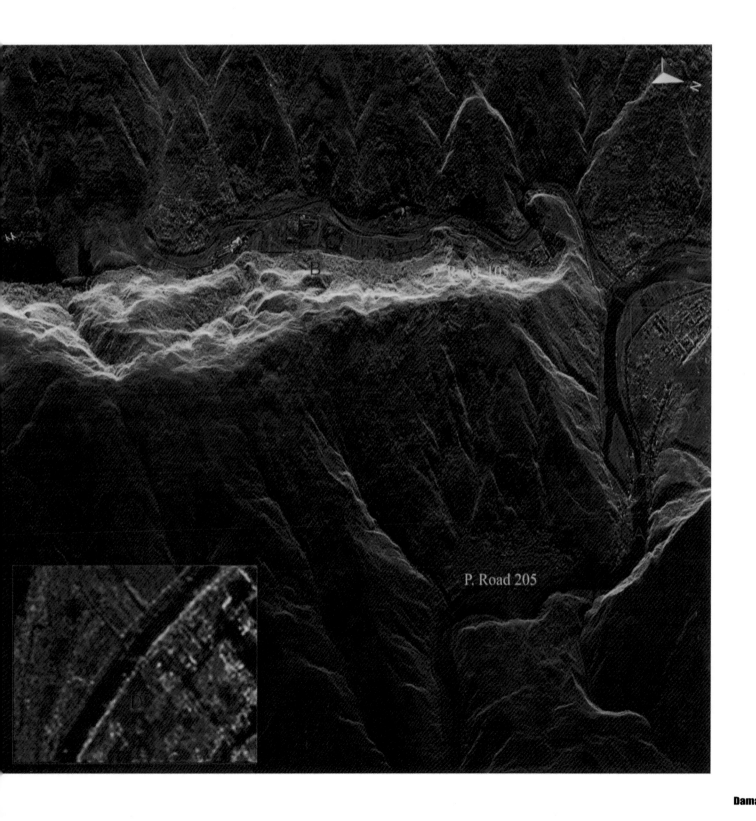

P. Road 205

▶ Photo of xiaoyudong Bridge after the earthquake

▼ **Airborne radar image over the town of Xiaoyudong, in the city of Pengzhou (May 17, 2008)**

This image shows the region from the town of Xiaoyudong to the town of Longmenshan that was affected by the earthquake and debris flow. Several parts of roads and railways through this region were ruined. (A) A road blocked by debris flow. (C) The collapsed Xiaoyudong Bridge (see the in situ photo), which lies in the northwest section of the city of Pengzhou.

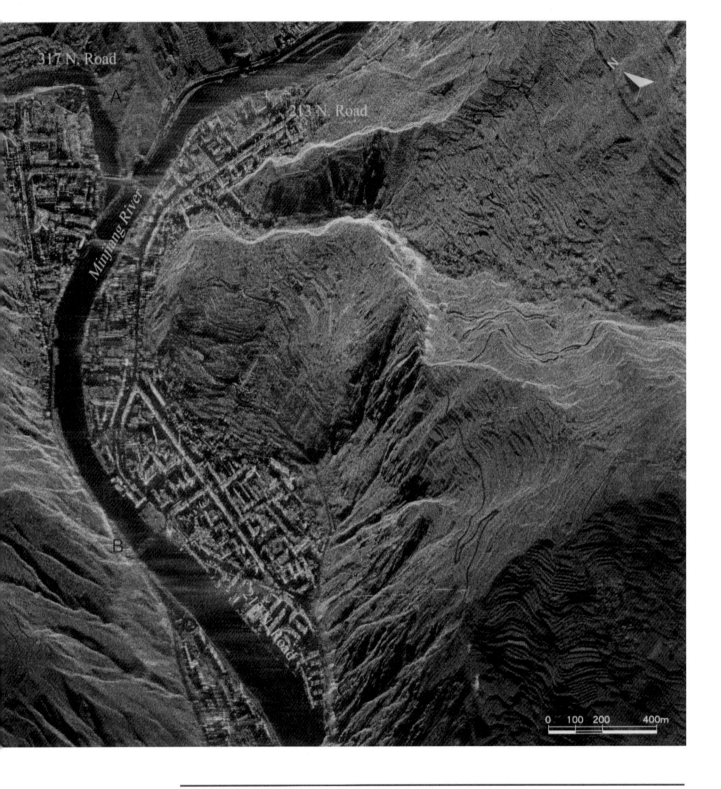

This image shows national highways G213 and G317 passing through Wenchuan County blocked by mountain landslides. (A) A damaged segment of G317, which leads to Lixian County. (B) Part of the highway that was blocked. (C) A portion of G213 that connects Wenchuan County with the city of Dujiangyan that was completely blocked by a huge landslide.

▶ Airborne radar image over the Zipingpu Reservoir (May 17, 2008)

This image shows roads in the north and south of the Zipingpu Reservoir that were blocked by landslides. (A) and (B) Roadways located in the town of Longchi north of the reservoir that were both ruined by the landslides. (C) and (D) Portions of national highway G213 that were also blocked, which is the only road leading to Wenchuan County and Maoxian County.

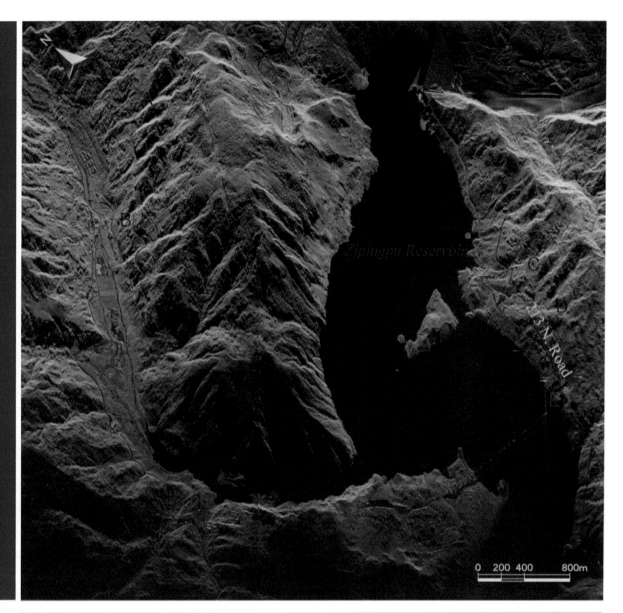

▶ *IKONOS* image over the urban district of Wenchuan County (May 18, 2008)

This image shows that many sections of national highways G213 and G317 through the county were destroyed by landslides triggered by the strong earthquake. (A) and (B) Segments of National Highway G213 were blocked. (C) A segment covered by debris. (D) and (E) Heavily damaged roadways. And the road from (F) to (G) was also blocked.

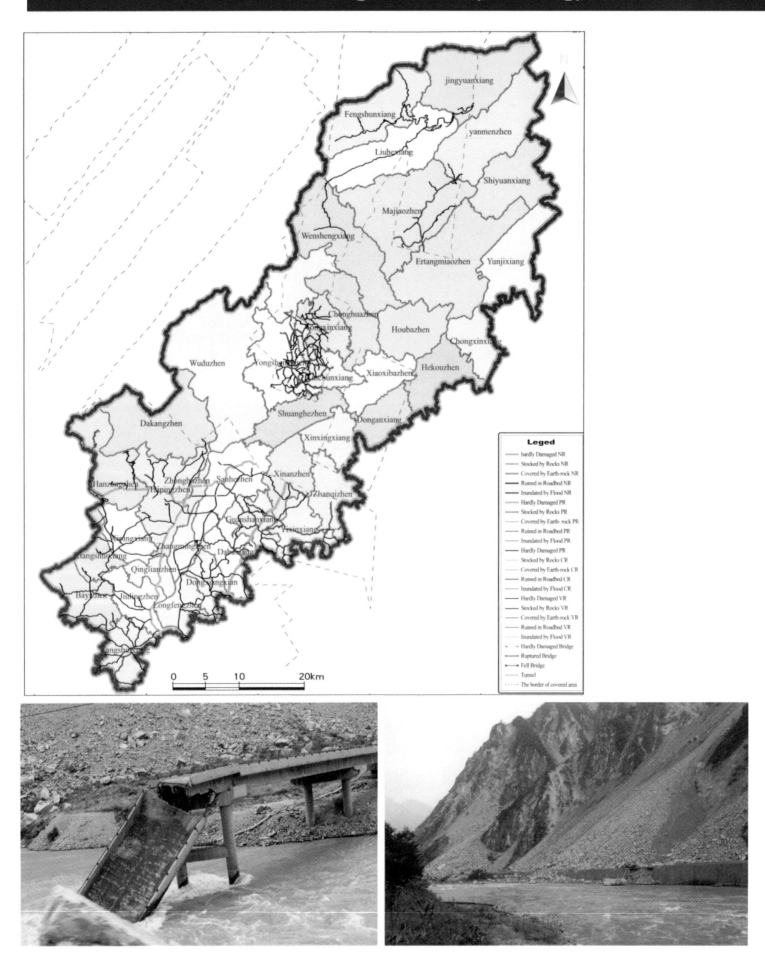

Leged

- Hardly Damaged NR
- Stocked by Rocks NR
- Covered by Earth-rock NR
- Ruined in Roadbed NR
- Inundated by Flood NR
- Hardly Damaged PR
- Stocked by Rocks PR
- Covered by Earth-rock PR
- Ruined in Roadbed PR
- Inundated by Flood PR
- Hardly Damaged PR
- Stocked by Rocks CR
- Covered by Earth-rock CR
- Ruined in Roadbed CR
- Inundated by Flood CR
- Hardly Damaged VR
- Stocked by Rocks VR
- Covered by Earth-rock VR
- Ruined in Roadbed VR
- Inundated by Flood VR
- Hardly Damaged Bridge
- Ruptured Bridge
- Fell Bridge
- Tunnel
- The border of covered area

0   5   10   20km

Road damage in Qingchuan County

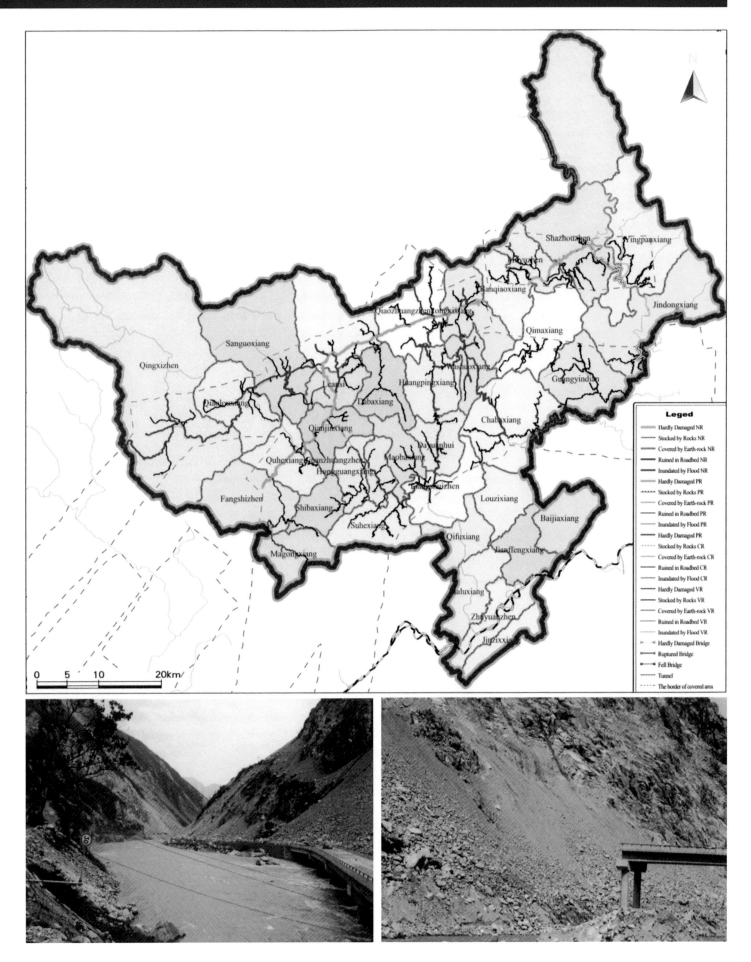

Leged

- Hardly Damaged NR
- Stocked by Rocks NR
- Covered by Earth-rock NR
- Ruined in Roadbed NR
- Inundated by Flood NR
- Hardly Damaged PR
- Stocked by Rocks PR
- Covered by Earth-rock PR
- Ruined in Roadbed PR
- Inundated by Flood PR
- Hardly Damaged PR
- Stocked by Rocks CR
- Covered by Earth-rock CR
- Ruined in Roadbed CR
- Inundated by Flood CR
- Hardly Damaged VR
- Stocked by Rocks VR
- Covered by Earth-rock VR
- Ruined in Roadbed VR
- Inundated by Flood VR
- Hardly Damaged Bridge
- Ruptured Bridge
- Fell Bridge
- Tunnel
- The border of covered area

Chapter

Destroyed Farmlands and Forests

Most of the farmlands and forests that were destroyed in the Wenchuan earthquake were located in the mountainous areas of the northwest Sichuan Basin in the Minshan Mountain Range between 2000 and 5000 m altitude. In this region, there are high mountains and many rivers, the main rivers being the Minjiang, Jianjiang, and Fujiang. Most of the river valleys are V shaped. This region is covered with high-density forests that contain a wide variety of trees and plants, encompassing more than 4000 species. The forest cover primarily comprises dense trees and shrubs, while sparse forest, younger forests, and reforested clear-cut areas make up less than 20% of the total forest. Larger areas of farmlands distributed in the mountains and hills of the earthquake region are mainly terraced. Moreover, local and national governments have established more than 20 natural conservation areas for giant panda protection in this region. About 40% of giant pandas around the country live in these conservation areas.

Many geological disasters, such as landslides, avalanches, rock debris flows, and mud-rock flows were triggered by aftershocks of the Wenchuan earthquake. These secondary geological disasters directly caused widespread ecological destruction of vegetation and farmlands. The secondary geological disasters are mainly distributed along human activity areas such as roads and rivers, deeply incised valleys that lie mainly along the rivers Minjiang, Jianjiang, Fujiang, and their tributaries. The cultivated environments around these farmlands were severely damaged, which will make recovery of cultivation very difficult. Landslides, avalanches, rock debris flow, and mud-rock flow present different characteristics of spectral and textural structure in the remote sensing images, so we can study changes of spectrum and texture to analyze and assess the degree of damage to the ecological environments of vegetation and farmlands.

This chapter presents various remote sensing images of destroyed farmlands, vegetation landscape, and panda habitats that are interpreted and analyzed using remote sensing satellite and aerial images collected from May 14 to May 28, 2008. The counties that sustained the most damage to farmlands and vegetation were Wenchuan and Beichuan counties.

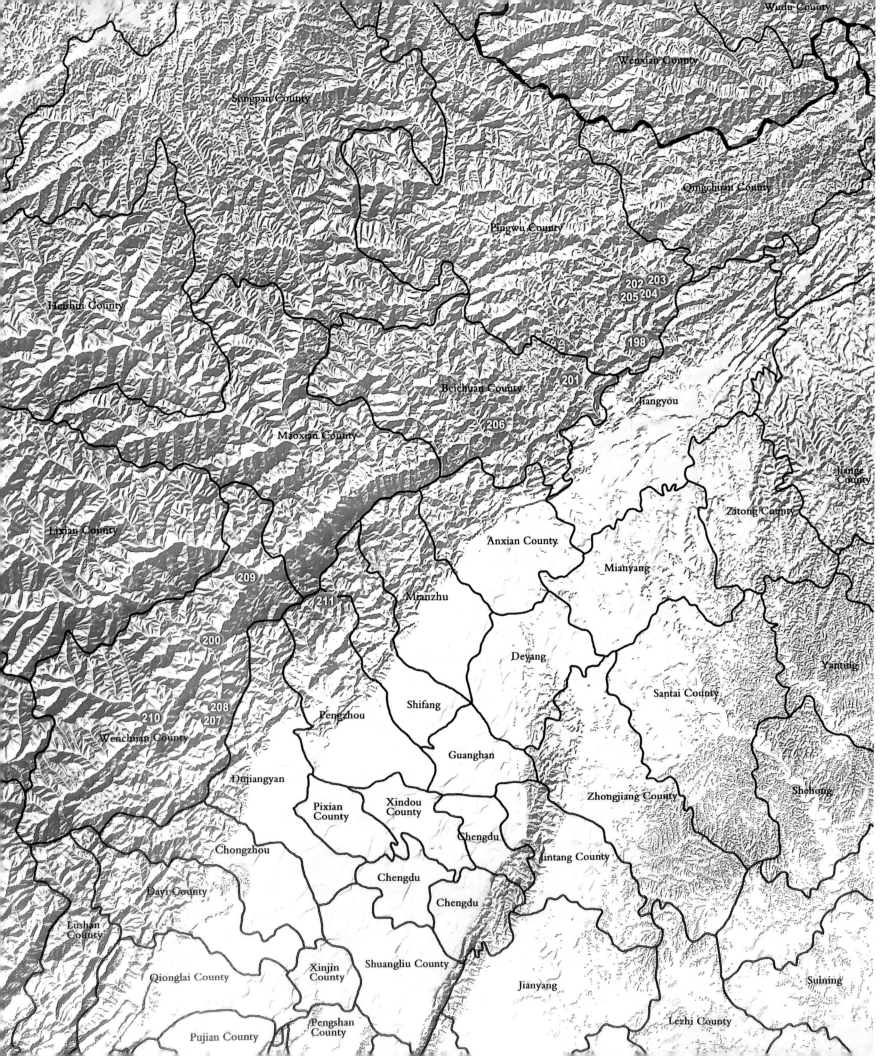

The beautiful terraces in Sichuan Province

The terraces that are cultivated in Sichuan Province were developed during the Tang Dynasty over 1000 years ago. These terraces make agricultural development in the mountainous area possible. However, after the May 12 Wenchuan earthquake, many terraces were destroyed by secondary geological disasters such as landslides, mud-rock flows, rock debris flow, barrier lakes, and so on.

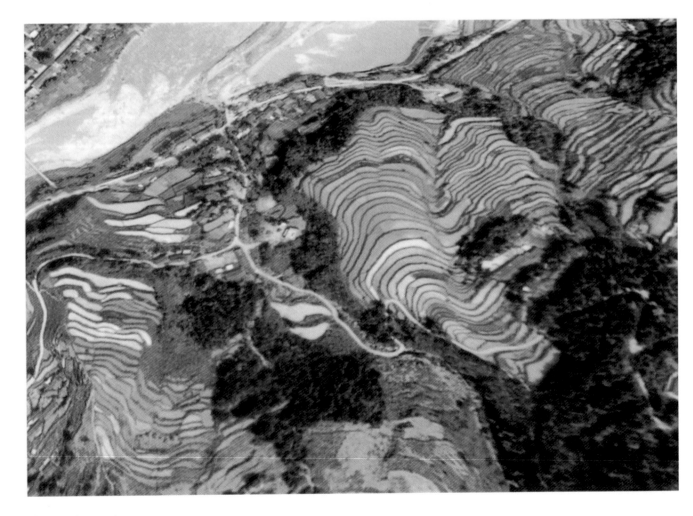

Distribution of destroyed farmlands in counties hit by the Wenchuan earthquake

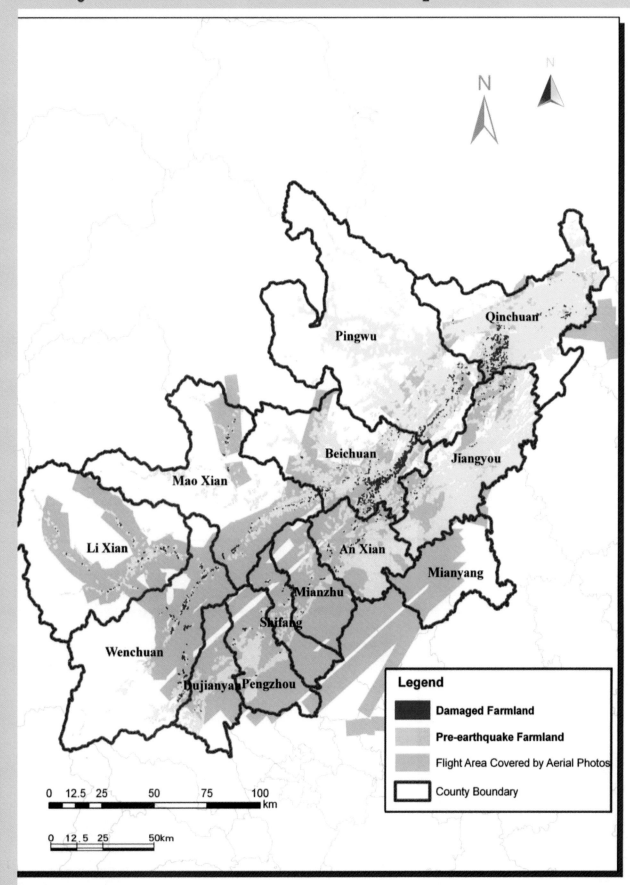

Legend

- Damaged Farmland
- Pre-earthquake Farmland
- Flight Area Covered by Aerial Photos
- County Boundary

This map shows the distribution of destroyed farmlands in 13 hard-hit counties in the May 12 Wenchuan earthquake. We determined which counties were most affected by analyzing airborne images acquired from May 14 to May 28, 2008. This analysis indicated that Wenchuan, Beichuan, and Qingchuan counties suffered the greatest damages.

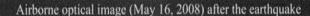

The terraces show gray or light-green colors, and the downslope area shows green color in the *SPOT* image on 9 May 2007. The farmlands that were buried or submerged by landslides, avalanches, or floods during the earthquake appear as white or gray colors in the airborne optical image on May 16, 2008. The earthquake not only destroyed farmlands, but also changed the cultivation environment. (A), (B), and (C) marked in the image (right) indicate the locations of destroyed terraces.

SPOT 5 image (May 9, 2007) before the earthquake

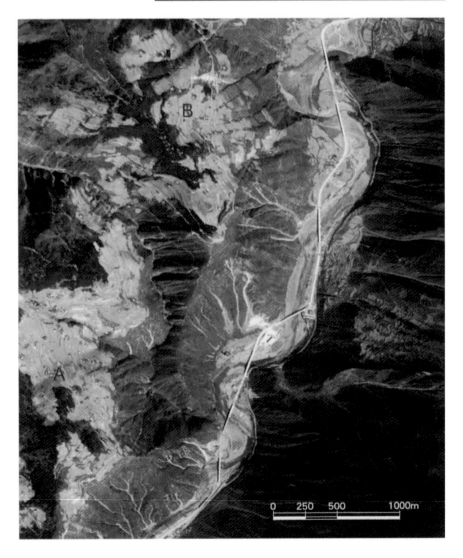

Airborne optical image (May 16, 2008) after the earthquake

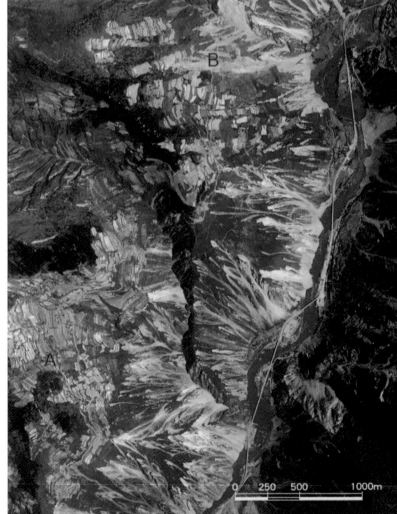

▲ **Airborne optical remote sensing image of the town of Xupingba**

Data were acquired on May 28, 2008. (A), (B), (C), (D), and (E) indicate destroyed terraces, and the red frame shows the location of the close-up image in the lower right-hand corner.

This *SPOT 5* image was acquired on May 16, 2008. Secondary geological disasters such as landslides, mud-rock flows, and so on (shown as the yellow or yellow-black in the image), in the town of Nanba and the town of Shuiguan, caused many barrier lakes to appear. (I), (II), and (III) marked in this image indicate the locations of the following three airborne color images.

▼ Image I: airborne optical remote sensing image acquired on May 28, 2008 around the village of Tangjiaba, in Pingwu County

(A), (B), (C), and (D) indicate the locations of destroyed farmlands

▲ Image II: airborne optical remote sensing image acquired on May 28, 2008 of the village of Liangjiasan, in Pingwu County

(A), (B), (C), (D), (E), (F), and (G) indicate the locations of destroyed farmlands.

(A), (B), (C), (D), (E), (F), and (G) indicate the locations of destroyed farmlands.

Town of Xuanping, May 27, 2008

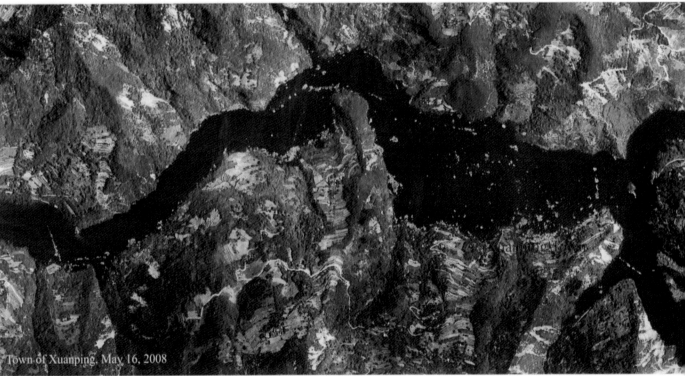

Town of Xuanping, May 16, 2008

▲ The inundated farmlands of Beichuan County

The two images above were acquired on May 16, 2008 and May 28, 2008, respectively, after the earthquake. Comparing these two images shows that many farmlands and houses of the town of Xuanping in the upper reaches of the Jianjian River were inundated by water, which was caused by the barrier lakes of Tangjiashan in Beichuan County.

Destroyed landscape in the Minjiang River Valley

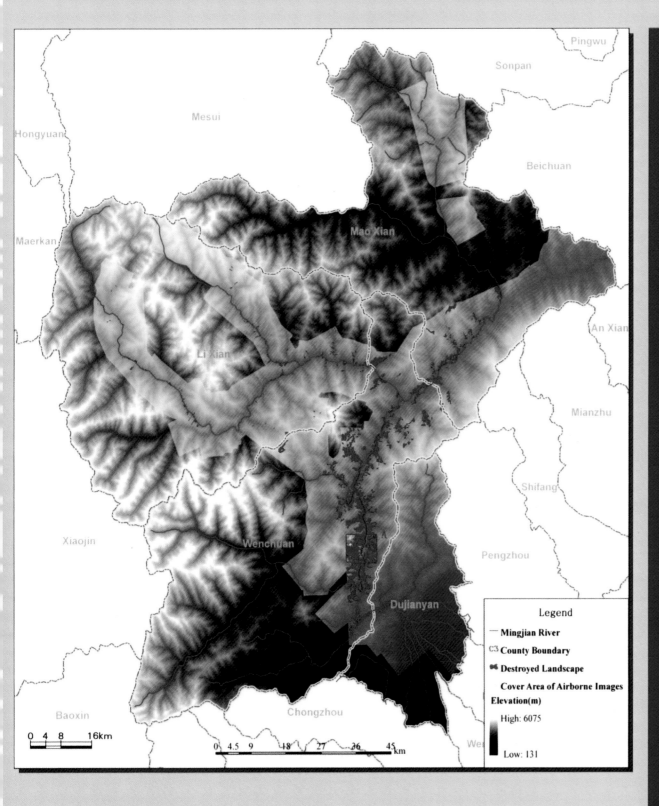

Legend
— Mingjian River
⊂⊃ County Boundary
⋈ Destroyed Landscape
    Cover Area of Airborne Images
Elevation(m)
    High: 6075
    Low: 131

0 4 8 16km

0 4.5 9 18 27 36 45 km

This is a composite remote sensing image of destroyed landscape that was created by overlaying digital terrain elevation generated from the airborne imagery onto an image of the area along the Minjiang River.

The image shows that the destroyed landscape is distributed along the valleys of the Minjiang, Jianjiang, and Fujiang rivers and their tributaries. The landscape destruction was caused by secondary geological disasters after the earthquake, including landslides, avalanches, and rock debris flows. The landscape changes will lead to the changes in the local ecological environment. Assessment of the landscape changes after the earthquake can provide scientific support for developing environmental reconstruction programs during the disaster recovery efforts.

There is a red rectangle in the lower left marked (A), showing the location of the remote sensing image in the typical natural landscape changes shown on the next page.

Destroyed landscape caused by secondary geological disasters

In Wenchuan County, large areas of forests, farmlands, and grass were destroyed by secondary geological disasters that followed the earthquake. Comparing the *SPOT 5* images on May 9, 2007 (before the earthquake) with the airborne images of the same region on May 16, 2008 (after the earthquake), it is obvious that the Wenchuan earthquake annihilated the original surface vegetation.

The images below are three-dimensional natural landscapes derived from images and Digital Elevation Models (DEM) viewing to the west of Xingwenpin of Yinxing Township. Comparing the characteristics of these four images shows the extent of landscape destruction.

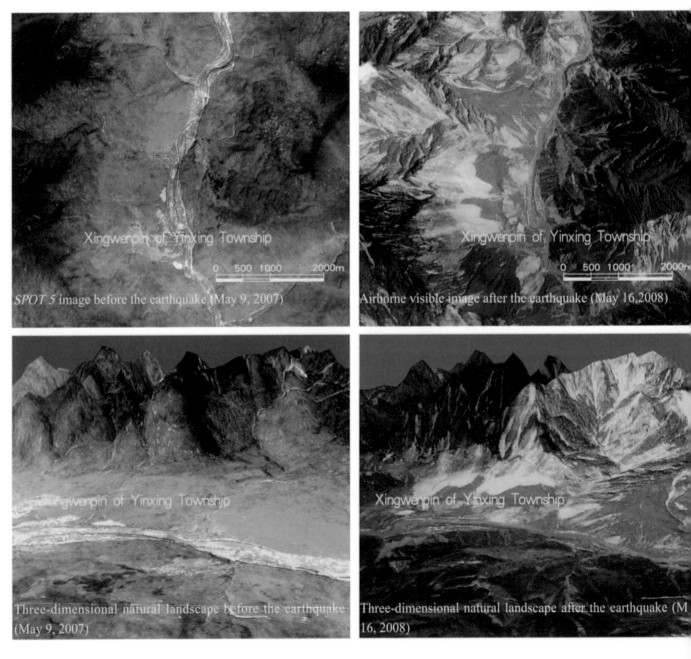

SPOT 5 image before the earthquake (May 9, 2007)

Airborne visible image after the earthquake (May 16, 2008)

Three-dimensional natural landscape before the earthquake (May 9, 2007)

Three-dimensional natural landscape after the earthquake (M 16, 2008)

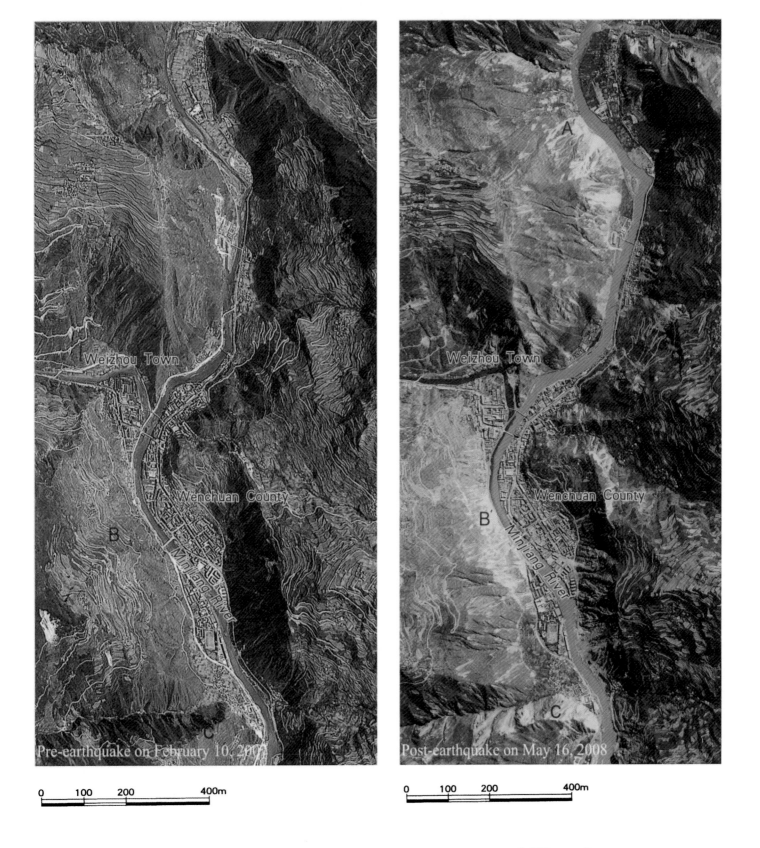

0 100 200   400m

0 100 200   400m

Secondary geological disasters around Wenchuan County

A *SPOT 5* image before the earthquake created on February 10, 2007 appears on the left. An airborne visible image from after the earthquake acquired on May 16, 2008 appears on the right. (A), (B), and (C) in the left-hand *SPOT* image and the same locations (A′), (B′), (C′) in the right image indicate the severely damage to the landscape caused by landslides, mud-rock flows, and rock debris flows.

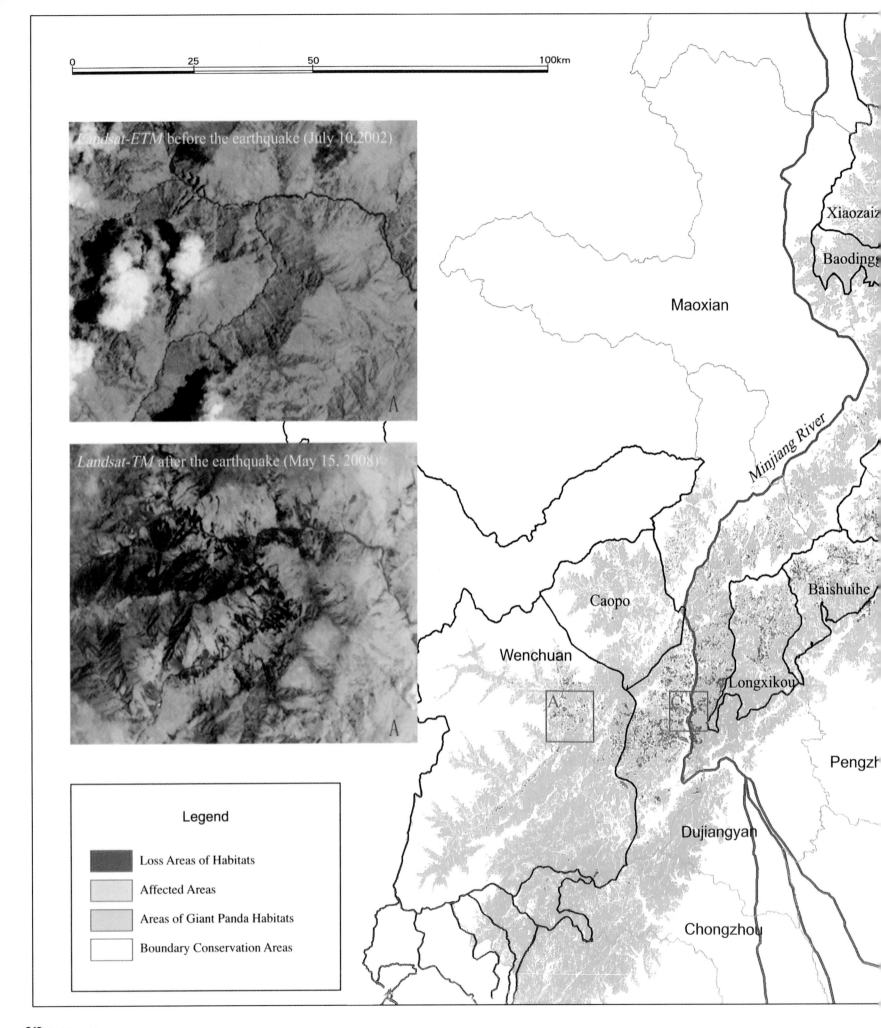

Landsat-ETM before the earthquake (July 10, 2002)

Landsat-TM after the earthquake (May 15, 2008)

Legend

Loss Areas of Habitats

Affected Areas

Areas of Giant Panda Habitats

Boundary Conservation Areas

Xiaozaiz

Baodingg

Maoxian

Minjiang River

Caopo

Baishuihe

Wenchuan

Longxikou

Pengzh

Dujiangyan

Chongzhou

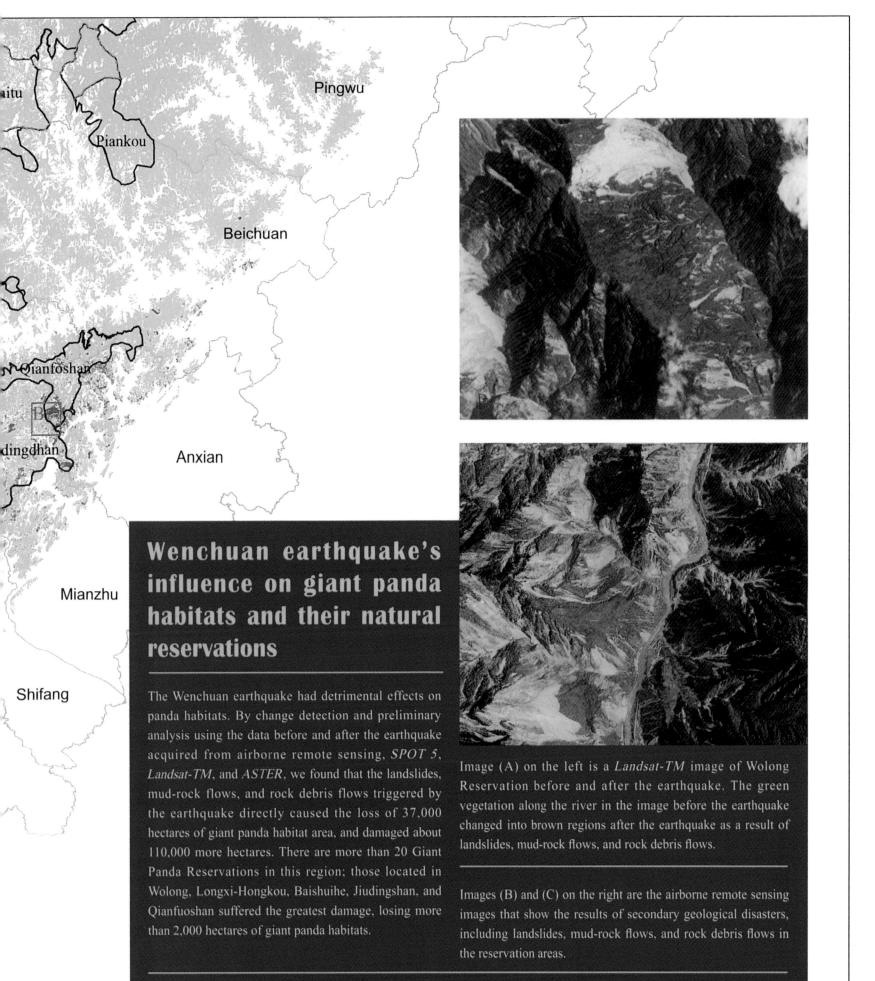

Pingwu

Piankou

Beichuan

Qianfoshan

Anxian

Mianzhu

Shifang

Wenchuan earthquake's influence on giant panda habitats and their natural reservations

The Wenchuan earthquake had detrimental effects on panda habitats. By change detection and preliminary analysis using the data before and after the earthquake acquired from airborne remote sensing, *SPOT 5*, *Landsat-TM*, and *ASTER*, we found that the landslides, mud-rock flows, and rock debris flows triggered by the earthquake directly caused the loss of 37,000 hectares of giant panda habitat area, and damaged about 110,000 more hectares. There are more than 20 Giant Panda Reservations in this region; those located in Wolong, Longxi-Hongkou, Baishuihe, Jiudingshan, and Qianfuoshan suffered the greatest damage, losing more than 2,000 hectares of giant panda habitats.

Image (A) on the left is a *Landsat-TM* image of Wolong Reservation before and after the earthquake. The green vegetation along the river in the image before the earthquake changed into brown regions after the earthquake as a result of landslides, mud-rock flows, and rock debris flows.

Images (B) and (C) on the right are the airborne remote sensing images that show the results of secondary geological disasters, including landslides, mud-rock flows, and rock debris flows in the reservation areas.

Chapter 7

DEMOLISHED INFRA-STRUCTURE

Not only did the earthquake cause significant losses in human lives and property, it also caused hidden trouble and damage to many infrastructures. After the Wenchuan earthquake, many important infrastructures in quake-hit areas were demolished, and hydrological engineering systems and mining construction areas were particularly badly damaged. Geological disasters, such as landslides, debris flows, and collapse, caused serious problems for these infrastructures. Some reservoir dams were buried by landslides, or damaged by debris flows, and the earthquake threatened dam security. Barrier lakes in the upper reaches of dam reservoirs filled some dams with mud and sands, and the dam system buildings were flooded and damaged, shutting down dam operations. Although some dams were not damaged because they were positioned on stable ground, the landslides, debris flows, and collapse that occurred near the reservoirs weakened the stability of the reservoirs. Moreover, some mining area construction faced severe losses because mines were blocked and mine buildings and houses were damaged.

Analysis of airborne and space-borne remote sensing images revealed many demolished infrastructures. This chapter shows the damage to important hydrological engineering systems and mining construction areas, including the Taipingyi power station in the town of Yinxing in Wenchuan County, the Shapai power station in the town of Caopo in Wenchuan County, the Futangba power station and Banpocun power station in the town of Miansi in Wenchuan County, the town of Nanxin power station in Maowen County, the Kuzhuba power station in the town of Qushan in Beichuan County, the Zipingpu Reservoir in the city of Dujiangyan, and the Jinhe Phosphate Mine in the town of Hongbai in the city of Shifang. Moreover, because radar images are sensitive to high-voltage wire transmission towers, we detected the towers affected by the landslides near National Highway 317 in western Wenchuan County and warned that the landslide could cause the towers to collapse and could interrupt power transmission.

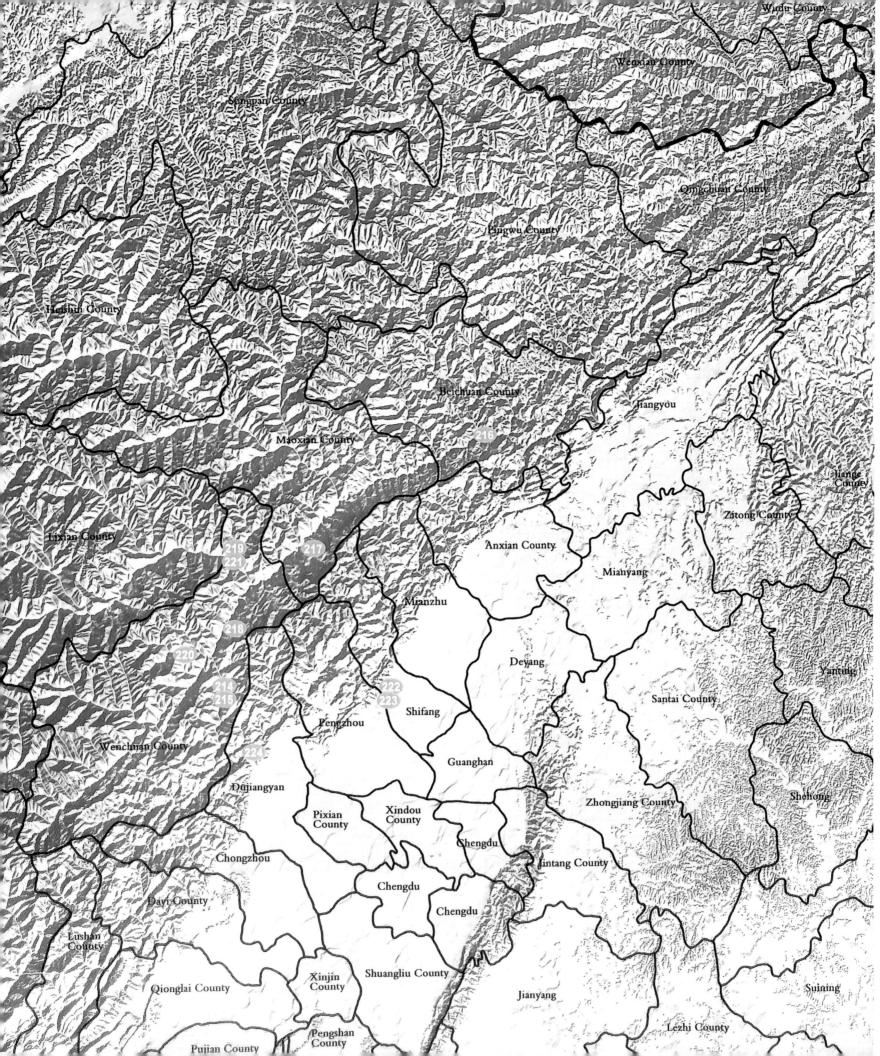

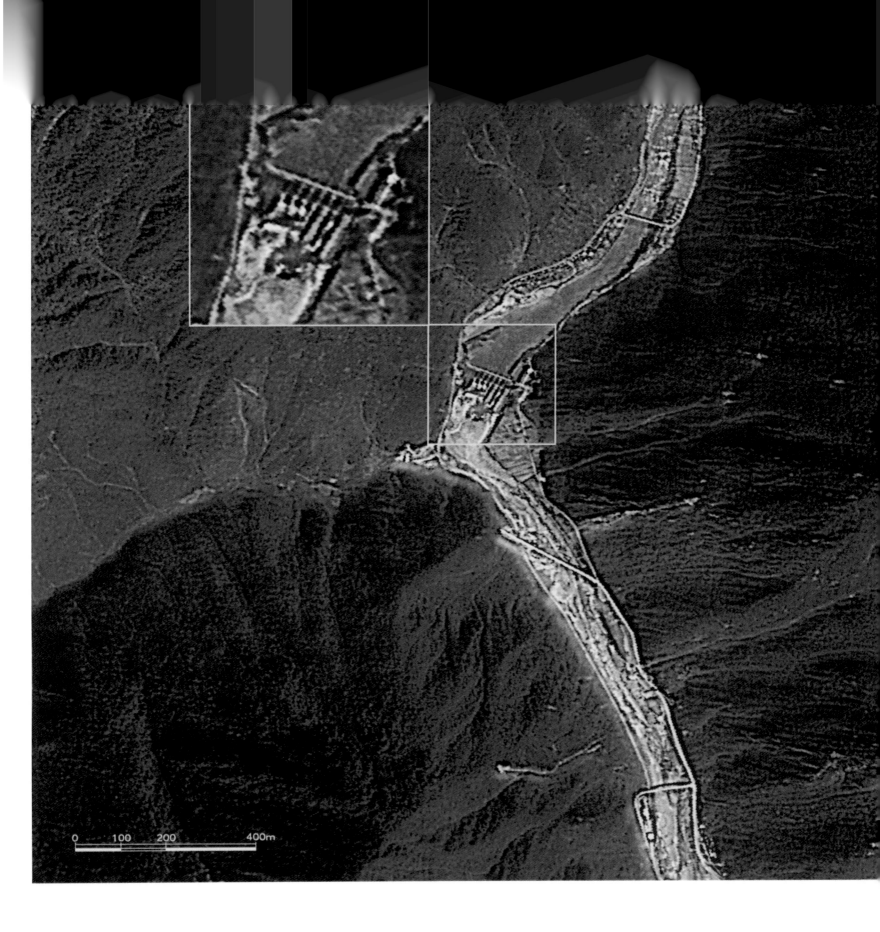

▲ Airborne optical remote sensing image of Taipingyi power station in the town of Yinxing, in Wenchuan County

---

This image, acquired on May 15, 2008, shows the Taipingyi power station, which was damaged by the landslides. A large landslide with a length of 530 m and a width of 500 m to the left of the station is threatening dam security.

◄ *SPOT 5* remote sensing image of the Taipingyi power station in the town of Yinxing, in Wenchuan County

---

This image, acquired on December 1, 2006, shows the Taipingyi power station and dam functioning properly before the earthquake.

▲ Airborne optical remote sensing image of the Kuzhuba power station in the town of Qushan, in Beichuan County

This image, acquired on May 16, 2008, shows the Kuzhuba power station affected by the barrier lake, Mud and sand filled up the dam, and the station buildings were flooded.

▶ Airborne optical remote sensing image of the power station in the town of Nanxin, in Maoxian County

This image, acquired on May 15, 2008, shows damage to the dam and the power station, which were inundated with water after the earthquake.

► Airborne optical remote sensing image of the Banpocun power station in the town of Miansi, in Wenchuan County

This image, acquired on May 17, 2008, shows that the dam of the Banpocun power station was badly affected by the earthquake. The west side of the dam was damaged by landslides.

▼ Airborne optical remote sensing image of the Futangba power station in the town of Miansi, in Wenchuan County

This image, acquired on May 17, 2008, shows that the Futangba power station was affected by the earthquake. The roads to the station were destroyed by landslides, and the buildings were also damaged.

This image, acquired on May 16, 2008, shows many high-voltage wire power transmission towers (A) located on the mountain beside National Highway 317 in western Wenchuan County. Some landslides (B, C) resulted from the earthquake. The location of a landslide that occurred several years ago on this mountain is marked (D); the earthquake produced new landslides, knocked over the towers, and damaged the power transmission equipment.

▼ Airborne optical remote sensing image of the Shapai power station in the town of Caopo in Wenchuan County

This image, acquired on May 17, 2008, shows that the dam of the Shapai power station has not been affected by the landslide and debris flow caused by the earthquake, but a large landslide occurred in the lower reaches of the dam.

Collapsing high-voltage wire
transmission towers

◄ Airborne optical remote sensing image of the Jinhe Phosphate Mine in the town of Hongbai, in the city of Shifang

This image, acquired on May 16, 2008, shows that the mining area was badly damaged. The mine was blocked off and the buildings and houses of the mining area collapsed because of the earthquake.

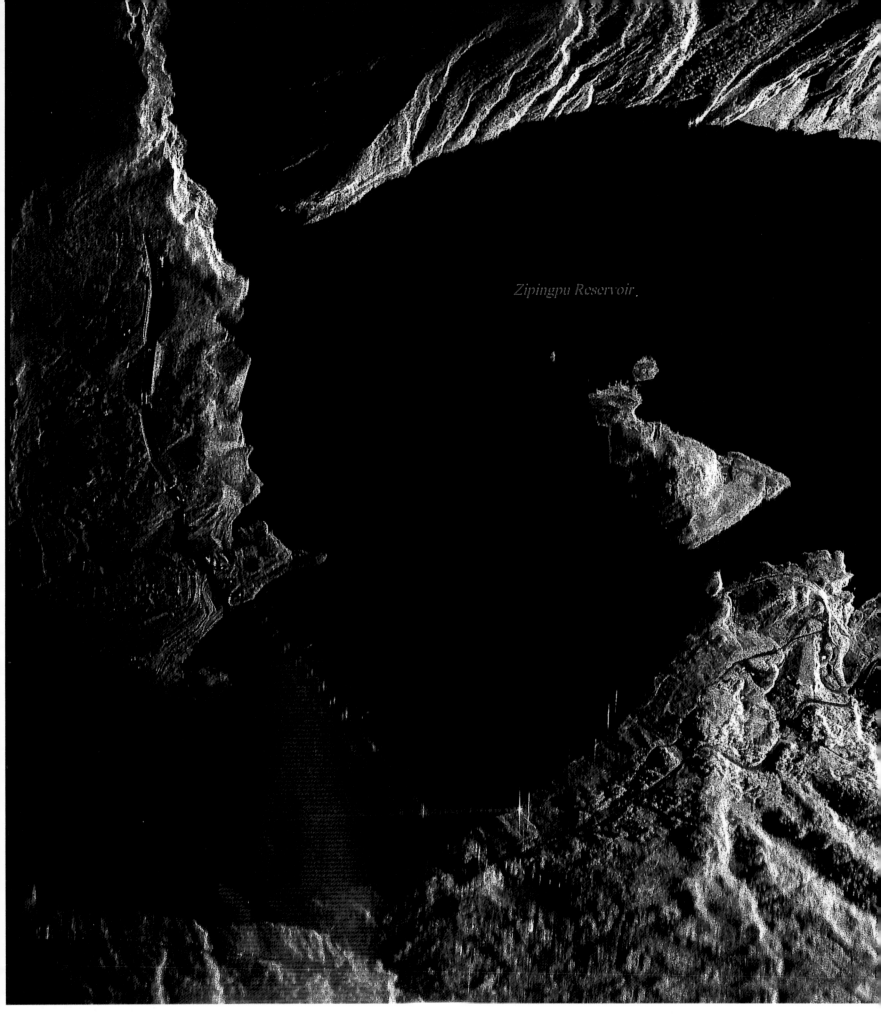

Zipingpu Reservoir.

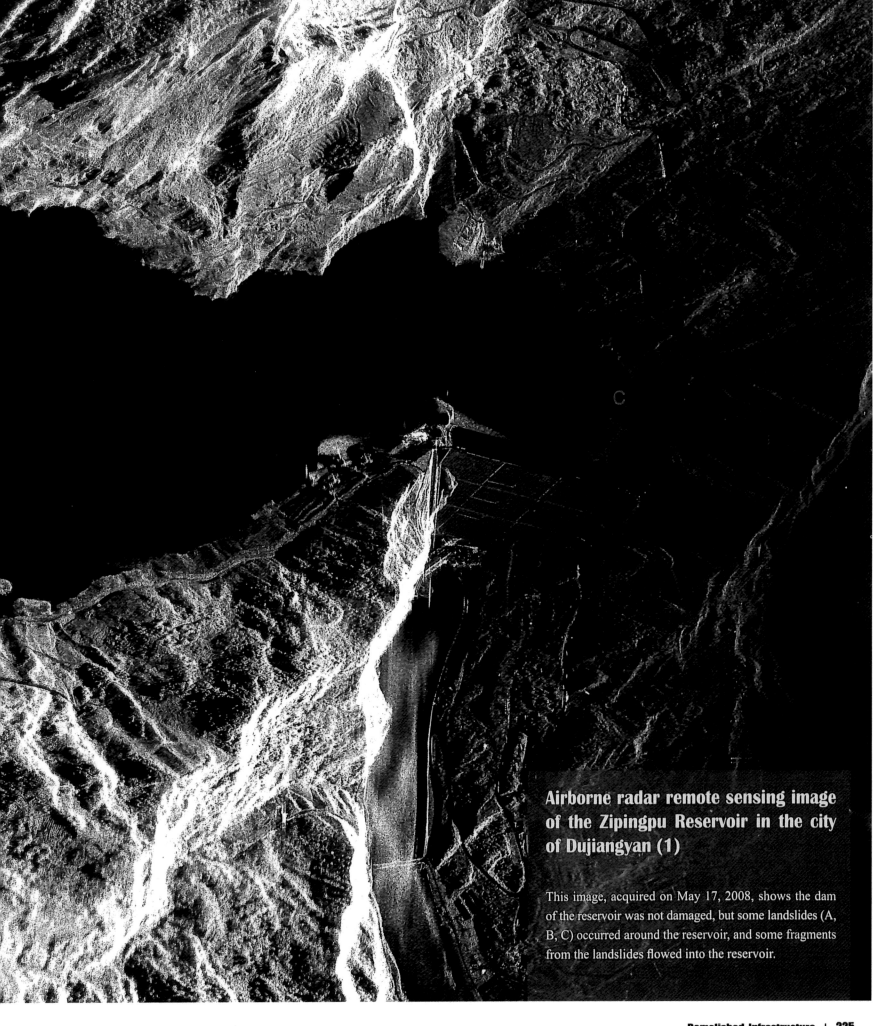

C

Airborne radar remote sensing image of the Zipingpu Reservoir in the city of Dujiangyan (1)

This image, acquired on May 17, 2008, shows the dam of the reservoir was not damaged, but some landslides (A, B, C) occurred around the reservoir, and some fragments from the landslides flowed into the reservoir.

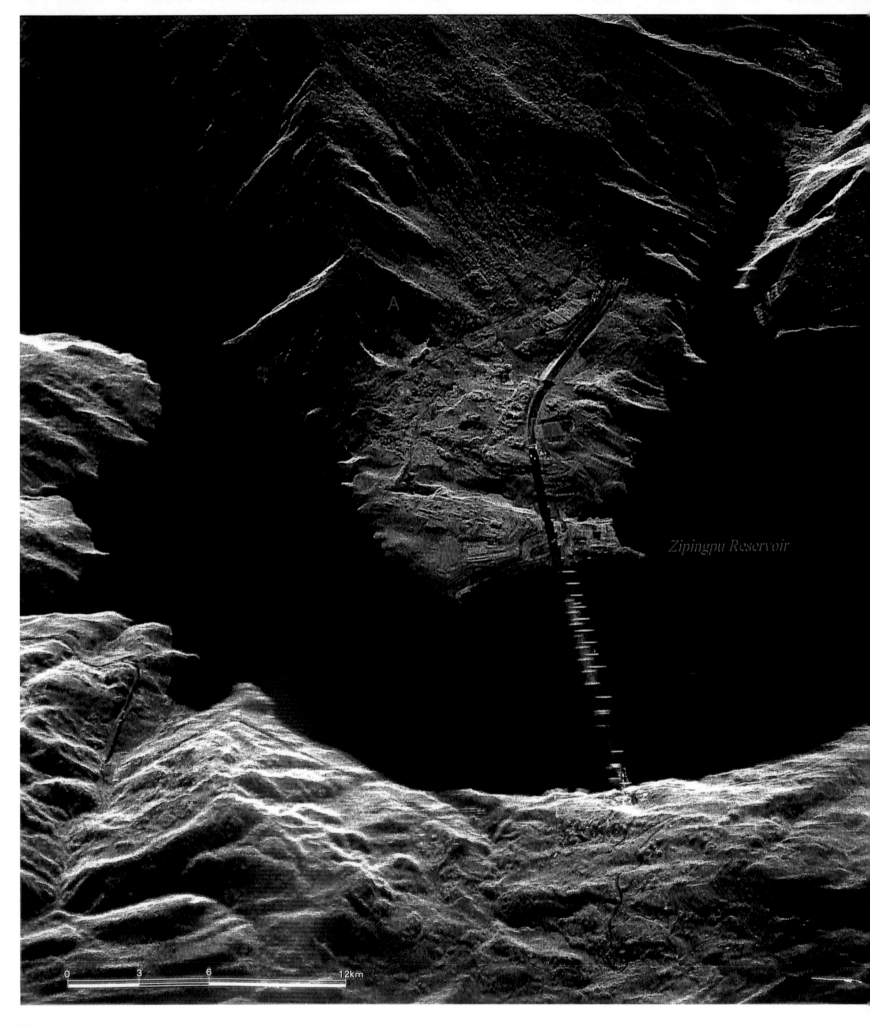

Zipingpu Reservoir

Airborne radar remote sensing image of the Zipingpu Reservoir in the city of Dujiangyan (2)

This image, acquired on May 16, 2008, shows some landslides (A, B, D) and collapse (C) that occurred around the Zipingpu Reservoir, and portions of landslides that flowed into the reservoir. The landslides are large and unstable.

Chapter

CIVILIZATION PERSEVERES

The Wenchuan earthquake-hit area includes many important national and cultural sites, including the world heritage Dujiangyan Irrigation Project, giant panda sanctuaries, Mount Qingcheng and the Dazu Rock Carvings, as well as homelands for ethnic minorities such as Tibetans, Qiangs, and Huis, and their cultural heritage sites. According to the statistics of the State Administration of Cultural Heritage of China (http://culture.people.com.cn/GB/87423/7235130.html), the Wenchuan earthquake affected area contains one site of world cultural heritage importance (the Dujiangyan Irrigation Project), 49 important heritage sites under state protection, 225 provincial-level cultural heritage sites, and 684 county-level cultural relic preservation areas.

After the earthquake, collapsed buildings in Dujiangyan were mainly reconstructions of ancient buildings (replicas of cultural relics); however, the genuine cultural relics were almost unaffected. While the buildings in the front part were partly destroyed, Mount Qingcheng as a whole was unaffected. The main hall of Erwang Temple, built during the Qing Dynasty, remains intact (http://news.sina.com.cn/pl/2008-05-30/135215650932.shtml). The Dujiangyan Irrigation Project is still solid and functions normally after the great earthquake, except for some fissures in the Yuzui Bypass Dike.

[1] http://culture.people.com.cn/GB/87423/7235130.html
[2] http://news.sina.com.cn/pl/2008-05-30/135215650932.shtml

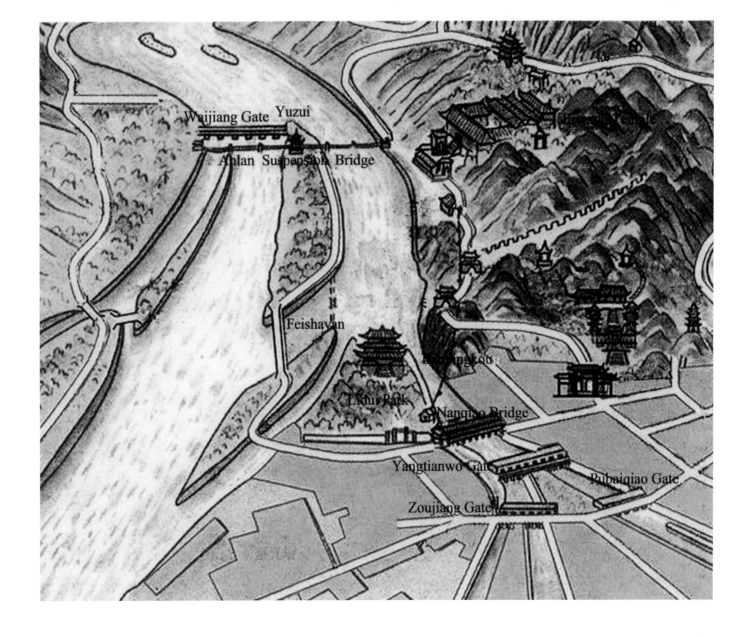

The Dujiangyan Irrigation System is the world's oldest and only remaining large hydraulic water project characterized by damless diversion. It consists of the Yuzui Bypass Dike, the Feishayan Floodgate, and the Baopingkou Diversion Passage. These three elements operate synchronously with one another to prevent flooding and to preserve the water supply to farmlands. The system automatically adjusts the capacity of the inner river for irrigation such that it carries 40% of upstream Minjiang water during flood seasons and 60% during dry seasons. Since the irrigation system was completed, the western Sichuan Plain has never flooded. This system has helped the Sichuan Plain earn the name "the Land of Abundance" in all seasons and all weather conditions.

The Yuzui Bypass Dike, named for its conical head said to resemble the mouth of a fish, is an artificial levee. The levee, which was built in the middle of the Minjiang River, functions to divide the river into inner and outer rivers. The outer river, situated in the west and known as "Jinma River." functions as the main stream and is mainly used to drain excess water and sand. The inner river, to the east along the foothills of Yulei mountain, is used for irrigation.

The Feishayan Floodgate, connecting the inner and outer rivers, primarily functions to control the flow of water. During the dry seasons it is not needed as much. When flooding occurs, it allows excess water to flow over the Feishayan Floodgate into the outer river. It can also help reduce the amount of silt and stone carried by the inner river before it flows into the Baopingkou.

The Baopingkou Diversion Passage, which Li Bing gouged through the mountain, is the final part of the system. Its name comes from the resemblance of Chengdu Plain to a large bottle and the passage's resemblance of the bottle neck. It works as a check gate to automatically control the volume of the water of the inner river.

Landsat 5-TM image of Dujiangyan City (acquired on June 26, 1994)

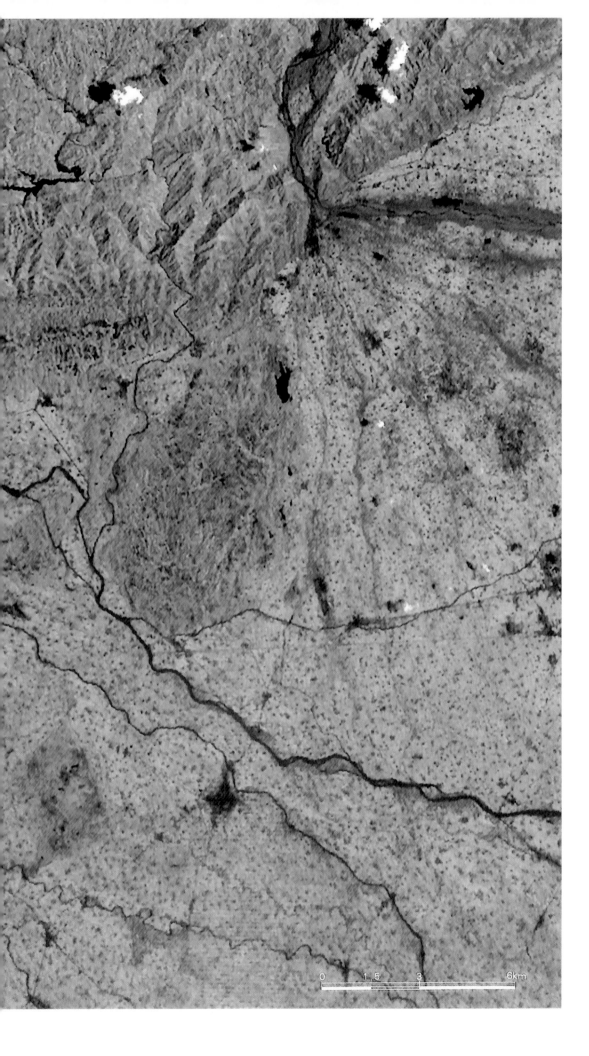

Landsat 5-TM image of Dujiangyan City (acquired on June 26th,1994)

Dujiangyan city is on the northwest edge of the Chengdu Plain, crossing the Longmenshan seismic belt in western Sichuan and the apex of the alluvial fan of the Minjiang River in the Chengdu Plain. The *Landsat 5-TM* image shows the topography gradually declining from northwest to southeast. With mountain in the northwest and plains in the southeast, the topography is ladder shaped. The proportion of mountains, plains, and water in the area is approximately 6:3:1, according to the saying "six parts mountain, three parts farmland and one part water." Designed by Li Bing, a governor of Shu Prefecture in the Kimdom of Qin during the Warring States period (over 2000 years ago), and his son, the world-famous Dujiangyan Irrigation System was erected on the Minjiang River at the foot of Yulei Mountain. Here the Minjiang River, after being joined by many tributaries in its upper reaches, runs toward the Chengdu Plain. With "digging deep for low dykes" as its motto, the Dujiangyan Irrigation System is characterized by damless diversion. After the Dujiangyan Irrigation System was built, the Minjiang River was divided into six canals for irrigation purposes, as well as the Jinma River, which carries the main stream. The whole system has been a crucial part of local agriculture for over 2000 years, and has contributed to the richness of Chengdu Plain and to earning its reputation as "The Land of Abundance." We are relieved that the Dujiangyan Irrigation Project was almost unaffected by the May 12 Wenchuan Earthquake.

SPOT 5 satellite image of the city of Dujiangyan (acquired on August 15, 2005)

IKONOS Satellite image (acquired on September 1

A. The Yuzui Bypass Dike

B. Waijiang Gate

C. Anlan Suspension Bridge

D. The Feishayan Floodgate

E. The Baopingkou Diversion Passage

F. Lidui Park

G. Nanqiao Bridge

H. Yangtianwo Gate

I. Pubai qiao Gate

J. Zoujiang Gate

K. Erwang Temple

Two main halls built during the Qing Dynasty remained mostly intact however, the front gate, the side halls, the slope protection works, and the rocky mountain paths, all built in 1990s, were badly damaged.

Airborne remote sensing image of Erwang Temple acquired after the earthquake (on May 18, 2008)

Picture of Erwang Temple after the earthquake

IKONOS satellite image acquired before the earthquake (on September 19, 2007)

Airborne remote sensing image acquired after the earthquake (on May 18, 2008)

The Baopingkou Diversion Passage and the Feishayan Floodgate were unaffected

The Dujiangyan Irrigation System is so astonishing not only because it is 200 years old and continues to function properly, providing irrigation to he Chengdu Plain, but also because its importance to the valley increases as esh water becomes a scarcer and more valuable resource. The Dujiangyan rigation System is also a unique architectural work, creating a model of armonious coexistence between mankind and nature. It makes man, earth, nd water a highly synchronous unity while making full use of rather than destroying natural resources, turning bane into boon, automatically dividing water, automatically removing silt, and automatically irrigating. The Dujiangyan Irrigation System represents a crystallization of ancient Chinese wisdom and an epoch-making masterpiece of Chinese civilization. It writes down a glorious page in the history of water conservancy in the world.

IKONOS satellite image acquired before the earthquake (on September 19, 2007)

Airborne remote sensing image acquired after the earthquake (on May 18, 2008)

Nanqiao Bridge, Liutianwo Gate, Pubaiqiao Gate, and Zoujiang Gate all remain intact.

Compared with the *IKONOS* satellite image acquired before the earthquake, this airborne remote sensing image acquired after the earthquake shows that the main parts of Dujiangyan Irrigation System are unaffected by the earthquake.

Airborne optical remote sensing image of Dujiangyan City (acquired on May 18, 2008)

A bell tower in Hanwang town, Mianzhu city, Sichuan Province, and the time stopped permanently at 14:28 on May 12, 2008.

INDEX

T - #0931 - 101024 - C412 - 280/248/19 - PB - 9781138112179 - Gloss Lamination